D1338110

WITHDRAWN

RTC Limerick

CLUSTER ANALYSIS ALGORITHMS
for data reduction and classification of objects

THE ELLIS HORWOOD SERIES IN COMPUTERS AND THEIR APPLICATIONS

Series Editor: BRIAN MEEK

Computer Unit, Queen Elizabeth College, University of London

The series aims to provide up-to-date and readable texts on the theory and practice of computing, with particular though not exclusive emphasis on computer applications. Preference is given in planning the series to new or developing areas, or to new approaches in established areas.

The books will usually be at the level of introductory or advanced undergraduate courses. In most cases they will be suitable as course texts, with their use in industrial and commercial fields always kept in mind. Together they will provide a valuable nucleus for a computing science library.

Published and in active publication

INTERACTIVE COMPUTER GRAPHICS IN SCIENCE TEACHING
Edited by J. McKENZIE, University College, London, L. ELTON, University of Surrey, R. LEWIS, Chelsea College, London.

INTRODUCTORY ALGOL 68 PROGRAMMING
D. F. BRAILSFORD and A. N. WALKER, University of Nottingham.

GUIDE TO GOOD PROGRAMMING PRACTICE
Edited by B. L. MEEK, Queen Elizabeth College, London and P. HEATH, Plymouth Polytechnic.

DYNAMIC REGRESSION: Theory and Algorithms
L. J. SLATER and H. M. PESARAN, Trinity College, Cambridge.

CLUSTER ANALYSIS ALGORITHMS: For Data Reduction and Classification of Objects
H. SPÄTH, Professor of Mathematics, Oldenburg University.

FOUNDATIONS OF PROGRAMMING WITH PASCAL
LAWRIE MOORE, Birkbeck College, London.

RECURSIVE FUNCTIONS IN COMPUTER SCIENCE
R. PETER, formerly Eotvos Lorand University of Budapest.

SOFTWARE ENGINEERING
K. GEWALD, G. HAAKE and W. PFADLER

PROGRAMMING LANGUAGE STANDARDISATION
Edited by B. L. MEEK, Queen Elizabeth College, London and I. D. HILL, Clinical Research Centre, Harrow.

FUNDAMENTALS OF COMPUTER LOGIC
D. HUTCHISON, University of Strathclyde.

SYSTEMS ANALYSIS AND DESIGN FOR COMPUTER APPLICATION
D. MILLINGTON, University of Strathclyde.

ADA: A PROGRAMMER'S CONVERSION COURSE
M. J. STRATFORD-COLLINS, U.S.A.

A COURSE IN COMPUTER SCIENCE
B. L. MEEK, Queen Elizabeth College, London, S. FAIRTHORNE, Queen Elizabeth College, London, P. WESTON, Bangor Technical College and P. MYERS, John Fisher School, Croydon.

CLUSTER ANALYSIS ALGORITHMS

for data reduction and classification of objects

HELMUTH SPÄTH
Professor of Mathematics
University of Oldenburg, Germany

Translated by Ursula Bull
Translation Editors: Owen Hanson and Brian Meek

ELLIS HORWOOD LIMITED
Publishers Chichester

Halsted Press: a division of
JOHN WILEY & SONS
New York - Chichester - Brisbane - Toronto

LIBRARY
LIMERICK COLLEGE
OF ART, COMMERCE
& TECHNOLOGY
class no: 519.53 SPA
acc.no: 3864

First published in 1980 by
ELLIS HORWOOD LIMITED
Market Cross House, Cooper Street, Chichester, West Sussex, PO19 1EB, England

The publisher's colophon is reproduced from James Gillison's drawing of the ancient Market Cross, Chichester.

Distributors:

Australia, New Zealand, South-east Asia:
Jacaranda-Wiley Ltd., Jacaranda Press,
JOHN WILEY & SONS INC.,
G.P.O. Box 859, Brisbane, Queensland 40001, Australia.

Canada:
JOHN WILEY & SONS CANADA LIMITED
22 Worcester Road, Rexdale, Ontario, Canada.

Europe, Africa:
JOHN WILEY & SONS LIMITED
Baffins Lane, Chichester, West Sussex, England.

North and South America and the rest of the world:
Halsted Press, a division of
JOHN WILEY & SONS
605 Third Avenue, New York, N.Y. 10016, U.S.A.

British Library Cataloguing in Publication Data
Späth, Helmuth
Cluster analysis algorithms. –
(Ellis Horwood series in computers and their applications).
1. Cluster analysis
2. Algorithms
I. Title II. Series
519.5'3 QA278 79-41653
ISBN 0-85312-141-9 (Ellis Horwood Ltd., Publishers)
ISBN 0-470-26946-4 (Halsted Press)

Typeset in Press Roman by Ellis Horwood Ltd.
Printed in Great Britain by Fakenham Press Ltd.

COPYRIGHT NOTICE
© Ellis Horwood Limited 1980
All Rights Reserved. No part of this publication may be reproduced, stored in a retrieval system, or transmitted, in any form or by any means, electronic, mechanical, photocopying, recording or otherwise, without the permission of Ellis Horwood Limited, Market Cross House, Cooper Street, Chichester, West Sussex, England.

Table of Contents

Author's Preface

The objective of cluster analysis is to separate a set of objects into constituent groups (classes, clumps, clusters) so that the members of any one group differ from one another as little as possible, according to a chosen criterion. The objects are each designated by a set of values of variables of (in general) different kinds. The purpose is to enable one to recognise, and to interpret in a plausible way, an existing structure of such a set of objects, to segment the objects and thus to achieve a data reduction, or to extract a hypothesis or basis for future prediction.

Cluster analysis can be of use in almost all of the empirical sciences. Indeed, very diverse procedures have been used successfully, for varying periods, in such fields as psychology, anthropology, medicine, criminology, biology, geology, archaeology and palaeontology, as well as in the social sciences, engineering, computer science, and the economic sciences (especially market research).

This book presents and explains the most important algorithms used in cluster analysis, and the problems associated with their use. In most cases, immediately applicable programs corresponding to the algorithms are also given (in Fortran, which is the most widely used scientific programming language), with worked examples illustrating their use. The mathematical knowledge needed to understand the algorithms – mainly of linear algebra – is minimal, so that this book will, it is hoped, promote the wider use of cluster analysis both in existing and in new fields of application.

Oldenburg, Spring 1975 Helmuth Späth

Note to the second edition: Some printing errors have been corrected (programs are unchanged) and the bibliography has been brought up to date.

Oldenburg, Spring 1977 Helmuth Späth

Note to the English translation of the second edition: Section 3.5 has been added and an error in the subroutine HIERCL has been corrected, together with the resulting errors in tables and figures. The programming error was announced in a supplement to the German second edition. In the opinion of the author, the method given in section 3.5 is a substantial contribution to the book. An index has also been added.

Oldenburg, Autumn 1979 Helmuth Späth

Editorial note: The general aim in preparing this translation has been to keep as close to the literal meaning of the original text as is consistent with readability. However, in some places a freer rendering has been given where it was felt desirable to make the English more idiomatic. This has not been done where it would have entailed a change in the ordering of the main equations and formulae.

With only minor exceptions, the original notations and figures have been retained, to facilitate comparison, though some changes have been necessary to accommodate the changes mentioned in the author's note above. Thus original Figs. U14-U21 are now Figs. U15-U22; Fig. U18 (originally U17), Fig. 13.1 and Fig. B36 have been modified; and Figs. B38-B40 are now Figs. B40-B42 of the original, Fig. B41 is the original Fig. B44, Fig. B42 is a modified version of the original Fig. B46, and Figs. B44-B45 are Figs. B47-B48 from the original. Fig. B43 in this edition is new, as is Fig. U14.

The U and H series figures, containing Fortran text, have been reproduced where appropriate from the original, except for the comment lines, for which English versions have been substituted. We are grateful to Prof. Späth for kindly supplying the new material for section 3.5 in English and reading the translation.

CHAPTER 1

Introduction

We shall assume that we have been given a data matrix

$$X = (x_{ik}) \quad (i=1,\dots,m;\ k=1,\dots,l) \tag{1.1}$$

representing m objects such as people, animals, plants, towns, postal districts, municipalities, households, textiles, computers, motor cars, television programmes, firms, documents, languages, etc. The m rows then give for each object the values of l variables (columns) denoting various characteristics of these objects. If the objects are people, such variables could be, for instance, sex, hair colour, mathematical ability, and physical size.

In general, four types of variables can be differentiated (see Aaker 1971, Anderberg 1973, Green and Tull 1970, Sokal and Sneath 1963). A variable is described as a 'nominal variable' if each of its possible different states, the different categories which it can represent, can be given a numerical code. For example, a variable representing 'marital status' could take the whole number values 1 to 4 for the conditions single, married, divorced and widowed respectively. Without loss of generality, it is usually possible to assign the first positive integers to the various values of a nominal variable. If the nominal variable can assume one of only two states – for example, the sex of a person, or whether or not the person wears spectacles – then these are usually denoted by the digits 0 and 1. In such cases the term 'binary variable' is also used. Instead of numbers, one can alternatively take as a set of values whichever letters of the alphabet seem most appropriate.

If the states of a nominal variable which are represented by numerical values can be arranged in a meaningful order, the term 'ordinal variable' is used. Such a variable is illustrated by the example of the possible responses 'very much', 'much', 'useful', 'not very useful', 'unsuitable' being given the values 1 to 5 on a bipolar scale. However, since the difference between two successive values is not necessarily of the same size, it has no quantitative meaning.

Finally, if the differences between successive values of an ordinal variable are of equal size, and if real number values as well as integer values are permitted, the term 'interval-scaled variable' is used. An example of this is the measurement of temperature in degrees centigrade. 'Ten degrees' has the same meaning anywhere on the scale. However, with an interval-scaled variable no meaningful zero or base point exists. For instance, one cannot say that 40°C is twice as hot as 20°C. This situation does not change even if temperature is measured from 'absolute zero', i.e. from −273°C approximately.

If there is an absolute base or zero point on the measurement scale of an interval-scaled variable, as for example for length or weight, the term 'ratio-scaled variable' is used. It is, for example, meaningful to say that 160 cm is to 80 cm as 40 cm is to 20 cm.

The handling of interval- and ratio-scaled variables for the purposes of cluster analysis depends only on equal differences in different values having the same significance, and thus that the arithmetic mean and (particularly) Euclidean distances are applicable. For that reason we shall call them metric variables in the following pages.

In a data matrix (1.1), nominal, ordinal and metric data can appear mixed or unmixed. The objective of all so-called multivariate methods is to find relationships within the data matrix. Considering these methods, a distinction can be made between dependent methods, in which one or more variables are considered to be dependent and the rest as independent, and interdependent methods, where all variables are dealt with simultaneously and equally (see Aaker 1971). For interdependent methods for a data matrix (1.1), the term 'Q-technique' is used when the objects are examined, and 'R-technique' is used when the variables are examined (see Joyce and Channon 1966). R- and Q-techniques are frequently applied one after the other; on the other hand, there are also techniques in which objects and variables are considered simultaneously (see McCormick *et al.* 1972, Hartigan 1971). Multivariate methods are further differentiated according to their applicability to various types of data.

We now take a brief look at the known multivariate methods, in order to make clear the position within them of cluster analysis.

The best known method is regression analysis, which attempts – predominantly by the use of metric data – to represent a variable as a linear combination of one or more variables in an optimal way, in the sense that the sum of the squares of the differences from linearity is minimised. If orthogonal least squares are used (see Späth 1974), then in contrast to customary regression analysis, the linear relationship can be solved for every variable. In this case the term 'interdependent regression analysis' can be used.

In discriminant analysis (see Aaker 1971) the dependent variable is nominal and the independent variables are metric. The aim is to assign objects to groups that are already represented by coefficients of linear functions.

Next, canonical correlation attempts to determine linear relationships between groups of dependent and independent metric variables.

Factor analysis (see Aaker 1971, Harman 1970, Uberla 1971) is an interdependent R-technique which starts with a correlation matrix of (metric) variables. It tries to find, with minimal loss of information, 'factors', i.e. linear combinations of those variables which are nearly linearly dependent, and are correlated, and whose effects in the data matrix therefore strongly depend upon one another. A large set of variables can thereby be reduced to a smaller number of factors capable of expressing its properties. This method is often used to reduce the data matrix as a preliminary to performing cluster analysis, partly to save time in calculation, but also to facilitate interpretation (see Aaker 1971, Bock 1970, Freitag 1972). Note that this assumes the existence of linear relationships. In psychology and in market research ordinal data matrices are also used, even though the construction of correlation matrices presupposes metric data. Factor analysis is also occasionally employed as a Q-technique, that is to say as a clustering technique, but this application of the method is not recommended (see Fleiss and Zubin 1969).

Given a data matrix containing non-metric or mixed variables, it is seldom meaningful to use Euclidean (i.e. classical geometric) distances to measure the proximity of objects. In such cases distance measures are used for which the triangular inequality does not necessarily hold. Such distance measures may be given directly, instead of the data matrix (1.1). Where this has been done, the Q-technique of multidimensional scaling (Aaker 1971, Green and Tull 1970, Green and Carmone 1970, Green and Rao 1972, Shepard *et al.* 1972) can be used to find a metric data matrix (1.1) from such a distance matrix, and the number of variables entering it. This number is initially unknown, and the variables are chosen so that they are as significant as possible, and in such a way that the Euclidean distances of the objects under consideration best fit the given distance values. If this method is only required to reproduce the same ranking order of the given distance values which is obtained by ranking the objects in pairs, this is called non-metric multidimensional scaling (see Kruskal 1964a, 1964b). These methods can be used as a preliminary to the use of clustering methods which need a data matrix of metric variables.

Finally, cluster analysis, also known in biology as numerical taxonomy, comprises interdependent processes for classifying objects described by sets of values of several variables. (Occasionally, however, the methods are used to classify the variables instead, or both are classified together, depending on the objects and variables involved.) The application of clustering methods is especially appropriate if more than two variables have to be considered at the same time. Realistic values for the number of objects are roughly $m = 50$ to $m = 5000$, and for the variables roughly $l = 2$ to $l = 10$.

The various clustering methods in (Anderberg 1973, Bock 1974, Duda and Hart 1973, Fisher and Van Ness 1971, Lance and Williams 1967a, 1967b), can be classified according to different criteria (see Green and Tull 1970). Essentially, one distinguishes between heuristic segmentation methods (Chapters 3 and 5), partitioning methods with an objective function (Chapter 3) and hierarchical methods (Chapter 4), either where all objects are taken together as one cluster and succeeding clusters are produced by division of some or all of those so far produced, or where objects are united into clusters step by step. Distance functions of any kind (Chapter 2) can be used with all methods. Many methods are defined only by an algorithm, rather than by an objective function.

Alongside these and several other methods, in which all variables can be considered simultaneously, there is also a smaller number of important methods, discussed briefly in Chapter 5, in which clusters are produced by dealing with the variables one at a time, in turn (see Anderberg 1973, Freitag 1972, Morgan and Sonquist 1963). The method used is frequently determined by the size of m and the types of variable available.

Basic problems in cluster analysis are: selection of distance, selection of algorithm, the number of clusters to be formed (where this is taken into account), and the choice of variables, especially their scaling.

In all, one can say that cluster analysis consists of a range of techniques which are – or whose applications are – mathematically based to a greater or lesser extent, and which can be regarded as successful – and indeed they usually are successful – if a data matrix contains a structure which is made clear and meaningful through interpretation of the clustering produced. Primarily, what makes an application of cluster analysis successful is the significant practical interpretation of the clusters it produces. For this reason it frequently makes sense to apply various methods, one after another and independently. Nevertheless one is sometimes happy enough simply to obtain a reasonable subdivision of the objects.

An essential part of the raison d'être of this book is, we consider, the provision of Fortran subroutines for the clustering methods discussed, and of main programs calling them, together with worked examples of their application. We confine ourselves to essentials in explaining these programs. In the presentation of the Fortran programs, which are written in an uncomplicated way and are easy to read, we describe the essential subroutine parameters only. Since there are no dynamic arrays in Fortran we have, for the sake of simplicity, given the same maximum array dimensions in the main programs as in the subroutines; however, in the subroutines the intended array dimensions are stated in comment lines. Input and output in the main programs can easily be reconstructed from the context and from the few READ and WRITE statements. Careful study of the worked examples is recommended.

Just a few more remarks on the captioning of the figures: diagrams are labelled B*i*, tables T*i*, subroutines U*i*, main programs H*i*, and tables of results E*i*, or, where there are several examples, E*i.j*. (The lettering comes from the original German text – e.g. H stands for Hauptprogram – and has been retained to facilitate cross-referencing between that text and the translation.) The index letter *i* is independent. H programs and E-tables with the same *i* correspond to each other.

The bibliography is arranged under the name(s) of the author(s) and the year of publication; where further distinction is necessary, an alphabetic suffix is added.

CHAPTER 2

Distance and Similarity Functions

2.1 METRIC DATA AND SCALING

Almost all clustering techniques involve a process of measurement, either of the magnitude of the distance between two objects, or of the magnitude of their similarity to each other, where the objects are described by values of the variables in the data matrix (1.1). We shall first give formal definitions of measurements which are independent of the data type, and then discuss those which are important in practice with nominal, ordinal, metric or mixed data, for measures both of distance and of similarity. No programs are given here, as these are very simple to set up and, in any case, are mostly to be found in (Anderberg 1973).

Let U denote a finite or infinite set of elements, which will in due course comprise the objects described by the rows of the data matrix: let $\mathbb{R}$ be the set of real numbers, $\mathbb{R}^+$ that of positive real numbers and $\mathbb{R}^-$ that of negative real numbers.

A mapping d: $U \times U \rightarrow \mathbb{R}$ (which therefore assigns a real number to each pair of elements of U) is called a distance function if, for arbitrary $x,y \in U$, we have

$$d(x,y) \geqq d_0, \tag{2.1.1}$$

$$d(x,x) = d_0, \tag{2.1.2}$$

$$d(x,y) = d(y,x). \tag{2.1.3}$$

in which d_0 is an arbitrary (also negative) finite real number. The relations (2.1.1) and (2.1.2) mean that d becomes minimal if the pair consists of identical elements. Formula (2.1.3) expresses the essential property of symmetry.

A distance function defined in this way is called metric if, in addition, the following conditions hold:

$$\text{If } d(x,y) = d_0 \text{ then } x = y, \tag{2.1.4}$$

$$d(x,z) \leqq d(x,y) + d(y,z), \quad z \in U. \tag{2.1.5}$$

Here (2.1.4) means that the smallest value of the distance function always implies that the two elements are identical. The statement (2.1.5) corresponds to the familiar triangular inequality of Euclidean geometry.

If in addition d_0 is set to zero, we obtain the metric concept of functional analysis. If d_0 is negative, a metric d' can be obtained from any metric distance function d by means of the definition $d'(x,y) = d(x,y) - d_0$.

A similarity function s is a mapping $s: U \times U \to \mathbb{R}$ with the following properties:

$$s(x, y) \leqq s_0, \tag{2.1.6}$$

$$s(x, x) = s_0, \tag{2.1.7}$$

$$s(x, y) = s(y, x). \tag{2.1.8}$$

Here, s_0 is a real number. The distinction between d and s lies in the statements (2.1.1) and (2.1.6).

A similarity function s is called metric if the following conditions hold:

$$\text{If } s(x,y) = s_0 \text{ then } x = y \tag{2.1.9}$$

$$[s(x, y) + s(y, z)] \cdot s(x, z) \geqq s(x, y) \cdot s(y, z). \tag{2.1.10}$$

Here (2.1.9) corresponds to the proposition that maximum similarity can only be possessed by two identical elements. The relation (2.1.10) is defined so as to be analogous with (2.1.5), and hence there is no major distinction between distance functions and similarity functions.

Let us give the following statements without proof (see Bock 1970, Deichsel 1972, Fritsche 1973, Soergel 1967): If d is a (metric) distance function, that has only $\mathbb{R}^+$ or $\mathbb{R}^-$ as its range of values, then $1/d$ is a (metric) similarity function. Furthermore, if d assumes finite values only, $\max d - d$, $\sqrt{\max d - d}$ and $\max d - d^2$ are metric similarity functions, as also is $\exp(-d)$, this last not requiring the assumption of the finiteness of d. The analogous assertions with s exchanged for d are also valid, provided that $\exp(-d)$ is replaced by $-\ln(s)$.

From now on, therefore, if d is a (metric) distance function then so are $-\ln(1/d) = \ln(d)$ and $-\ln(\max d - d)$; and if s is a (metric) similarity function with $s_0 = 1$ then so is $1/(1 - s)$.

Furthermore, the following assertion (Majone and Sanday 1968) holds: if d is a metric distance function, then so is $Md/(1 + d)$, where M is an arbitrary positive constant. A proof can be found in (Anderberg 1973).

In what immediately follows we shall consider specific (metric) distance and similarity functions, firstly for metric and then for binary data. After this it will be possible to deduce measures for general nominal and ordinal data.

Let x and y denote two real vectors $(x_1, \ldots, x_l)^T$ and $(y_1, \ldots, y_l)^T$, which will be used to correspond to two objects described by rows of the data matrix.

The best known of all metrics, which corresponds to the generalisation to more than two dimensions of the distance between two points in a plane, is derived from the L_2-norm of a vector x; namely, from

$$\|x\|_2 = \sqrt{\sum_{k=1}^{l} x_k^2} = \sqrt{x^T x} \tag{2.1.11}$$

we obtain

$$d_2(x, y) = \|x - y\|_2 = \sqrt{(x-y)^T (x-y)}. \tag{2.1.12}$$

This Euclidean metric has, together with the norm (2.1.11), the property that all its values are invariant with respect to orthogonal mappings (rotations) of the vectors which are described by the $l \times l$ matrices Q such that $Q^T Q = I$ (where I is the identity matrix). Thus we also have

$$\|Q x\|_2 = \|x\|_2 \tag{2.1.13}$$

and

$$d_2(Q x, Q y) = d_2(x, y). \tag{2.1.14}$$

These orthogonal mappings are the only mappings other than translations $x \to x + a$ for which d_2 is invariant.

The Euclidean metric can be generalised in two ways. One is by the application of the L_p-norm

$$\|x\|_p = \sqrt[p]{\sum_{k=1}^{l} |x_k|^p} \quad (p \geqq 1), \tag{2.1.15}$$

from which one obtains, by analogy,

$$d_p(x, y) = \|x - y\|_p. \tag{2.1.16}$$

No general mappings other than translations are known for which the L_p-norm and the d_p metric are invariant. The inequality $d_p(x,y) \leqslant d_q(x,y)$ holds for all x,y if and only if $p \geqslant q$ (Duran and Odell 1974).

$p = 1$ and $p = \infty$ are special cases. The norms

$$\|x\|_1 = \sum_{k=1}^{l} |x_k| \quad \text{and} \quad \|x\|_\infty = \max_k |x_k| \tag{2.1.17}$$

have, corresponding to them, the metrics

$$d_1(x, y) = \|x-y\|_1 \quad \text{and} \quad d_\infty(x, y) = \|x-y\|_\infty. \tag{2.1.18}$$

The second kind of generalisation is obtained by defining

$$\|x\|_B = \sqrt{x^T B x}, \tag{2.1.19}$$

instead of (2.1.11). Here B is a positive definite matrix – i.e. a symmetric matrix such that $x^T Bx \geqslant 0$ for all x and $x^T Bx = 0$ if and only if $x = 0$.

The metric corresponding to (2.1.19) is then

$$d_B(x, y) = \sqrt{(x-y)^T B (x-y)}. \tag{2.1.20}$$

In simple cases B is a diagonal matrix, the diagonal elements of which are different positive weights for the components of the vectors which correspond to the variables in the data matrix (1.1). Orthogonal mappings that leave the L_2-norm invariant correspond here to matrices P for which $P^T BP = B$, which in their turn leave (2.1.19) and consequently (2.1.20) invariant under translations.

By specifying a suitable B for the data matrix (1.1), we get the so-called Mahalanobis metric for the objects, which has the most general invariance property, namely of being invariant under any non-singular transformation C.

To do this we first need the covariance matrix of the variables (columns) of x_{ik}. We denote the kth column and the ith row of the data matrix (1.1) by $x_{.k}$ and $x_{i.}$ respectively; the corresponding means are denoted by $\bar{x}_{.k}$ and $\bar{x}_{i.}$. The matrix $S = (s_{kj})$ defined by

$$s_{kj} = \frac{1}{m} \sum_{i=1}^{m} (x_{ik} - \bar{x}_{.k})(x_{ij} - \bar{x}_{.j}) \qquad (k, j = 1, \ldots, l) \tag{2.1.21}$$

is then called the covariance matrix of the variables. The covariance matrix T of the objects is defined analogously as

$$t_{ij} = \frac{1}{l} \sum_{k=1}^{l} (x_{ik} - \bar{x}_{i.})(x_{jk} - \bar{x}_{j.}) \qquad (i, j = 1, \ldots, m) \tag{2.1.22}$$

The corresponding correlation matrices, which are needed, among other things, for the R- and Q-techniques of factor analysis, are thus given by

$$r_{kj} = \frac{s_{kj}}{\sqrt{s_{kk}\, s_{jj}}} \quad (k, j = 1, \ldots, l) \tag{2.1.23}$$

and

$$\varrho_{ij} = \frac{t_{ij}}{\sqrt{t_{ii}\, t_{jj}}} \quad (i, j = 1, \ldots, m) \tag{2.1.24}$$

respectively.

If we write

$$\widetilde{X} = (\widetilde{x}_{ik}) = (x_{ik} - \overline{x}_{.k}) \tag{2.1.25}$$

the matrix S can be written in the following form:

$$S = \frac{1}{m} \widetilde{X}^T \widetilde{X}, \tag{2.1.26}$$

from which it can be seen that, when the columns of $\widetilde{X}$ and hence of X are linearly independent – as can generally be assumed for $m \gg l$ – the covariance matrix is positive definite. Consequently S is then non-singular, and the positive definite inverse S^{-1} exists.

Following (Bock 1974, Morrison 1967, Overall 1964),

$$d_S(x_{i.}^T, x_{j.}^T) = \sqrt{(x_{i.} - x_{j.}) S^{-1} (x_{i.} - x_{j.})^T} \tag{2.1.27}$$

is called the 'Mahalanobis distance' between the two objects in the data matrix (1.1). For the m objects, a metric is thus defined which is invariant under all non-singular transformations C ($l \times l$ matrices) applied to the row vectors. That is to say, if we write $y_{i.}^T = C^T x_{i.}^T$ such that $y_{i.} = x_{i.}\, C\, (i = 1, \ldots, m)$, then

$$\frac{1}{m} \widetilde{Y}^T \widetilde{Y} = \frac{1}{m} (\widetilde{X} C)^T \widetilde{X} C$$

and accordingly

$$d_S(y_{i.}^T, y_{j.}^T) = \sqrt{(x_{i.} - x_{j.})\, C \left[\frac{1}{m} (\widetilde{X} C)^T \widetilde{X} C\right]^{-1} C^T (x_{i.} - x_{j.})^T}$$
$$= d_S(x_{i.}^T, x_{j.}^T).$$

If in particular C is a diagonal matrix with non-zero diagonal elements, the transformations of X by C means that the value of each variable in X is multiplied by a constant. This is termed a C scaling, and the Mahalanobis metric is said to be scale invariant, since it provides the same measure of distance whatever unit is used to measure the variables. The other metrics, in particular the Euclidean ones, do not possess this remarkable property.

(Morrison 1967) further recommends that the variables be given positive weights, of any magnitude, chosen by the user, by introducing a diagonal matrix D into (2.1.27), as follows:

$$d_{S'}(x_{i.}^T, x_{j.}^T) = \sqrt{(x_{i.} - x_{j.})\, D\, S^{-1}\, D(x_{i.} - x_{j.})^T} \tag{2.1.28}$$

In applications of clustering techniques, the Mahalanobis metric suffers from the disadvantage that the matrix is based on all the objects together and not, as would perhaps be more meaningful, separately on the objects in each cluster which are still unknown (Späth, Müller 1979); furthermore, its calculation is necessarily much more burdensome than that of the other metrics. For this reason, we have not used it in Chapter 3.

Moreover, if the objects are normalised so that $\|x_{i.}^T\|_2 = 1$ $(i=1, \ldots, m)$, it follows that, since S degenerates to the unit matrix, $d_S(x_{i.}^T, x_{j.}^T) = d_2(x_{i.}^T, x_{j.}^T)$, i.e. after this transformation the Mahalanobis metric and the Euclidean metric are identical.

The normalisation mentioned above has a further effect upon the Euclidean metric. It becomes, in fact,

$$\begin{aligned} d_2^2(x_{i.}^T, x_{j.}^T) &= \|x_{i.}^T - x_{j.}^T\|_2^2 = \|x_{i.}^T\|_2^2 + \|x_{j.}^T\|_2^2 - 2\, x_{i.}\, x_{j.}^T \\ &= 2(1 - \varrho_{ij}), \end{aligned} \tag{2.1.29}$$

in which ϱ_{ij} is the correlation matrix defined in (2.1.24).

Since we have obtained, from the correlation matrix of the objects (2.1.24), a (non-metric) similarity function – though its application is of doubtful value (see Fleiss and Zubin 1969, Kendall 1966) – the result (2.1.29) can be regarded as giving a relationship between distance and similarity functions, as discussed earlier.

Before going on to discuss other distance measures, we wish to examine normalised measure scaling in greater detail. By normalisation of the objects is meant the use of the transformation

$$x'_{ik} = \frac{1}{s_i}(x_{ik} - \bar{x}_{i.}) \quad \text{where} \tag{2.1.30}$$

$$s_i^2 = \sum_{k=1}^{l} (x_{ik} - \bar{x}_{i.})^2, \tag{2.1.31}$$

which yields

$$\|x_{i.}'^T\|_2 = 1 \tag{2.1.32}$$

and by normalisation of the variables is meant the use of the transformation

$$x''_{ik} = \frac{1}{t_k}(x_{ik} - \bar{x}_{.k}) \quad \text{where} \tag{2.1.33}$$

$$t_k^2 = \sum_{i=1}^{m} (x_{ik} - \bar{x}_{.k})^2, \tag{2.1.34}$$

which yields

$$\|x''_{.k}\|_2 = 1 \tag{2.1.35}$$

In the subroutine TRANSF (Fig. U1) the columns of a matrix X(M,L) are so transformed that their standard deviations

$$s_k = \sqrt{\frac{1}{m-1}\sum_{k=1}^{l}(x_{ik} - \bar{x}_{i.})^2} = \sqrt{\frac{1}{m-1}\left[\sum_{k=1}^{l} x_{ik}^2 - \frac{1}{m}\left(\sum_{k=1}^{l} x_{ik}\right)^2\right]} \tag{2.1.36}$$

are equal to 1 and their means are equal to 0. For a clustering technique based on the familiar Euclidean metric, (Bock 1970, Sokal and Sneath 1963) recommend that this transformation be performed before the application of the clustering algorithm – all variables, in other words, should be measured on the same scale. This, like other kinds of scaling, is suspect if there are substantial variations between the standard deviations.

```
      SUBROUTINE TRANSF (M,L,X)
C
C     EACH OF THE L COLUMNS OF THE MATRIX X(I,K)
C     (I = 1,....,M; K = 1,....,L) IS TRANSFORMED SO
C     THAT THE MEAN OF ITS ELEMENTS IS EQUAL TO ZERO
C     AND ITS STANDARD DEVIATION IS EQUAL TO ONE.
C
C     DIMENSION X(M,L)
      DIMENSION X(100,12)
      F=FLOAT(M)
      DO 3 K=1,L
            T=0.
            U=0.
            DO 1 I=1,M
                  V=X(I,K)
                  T=T+V
                  U=U+V*V
    1       CONTINUE
            Q=T/F
            S=SQRT((F-1.)/(U-T*Q))
            DO 2 I=1,M
                  X(I,K)=S*(X(I,K)-Q)
    2       CONTINUE
    3 CONTINUE
      RETURN
      END
```

Fig. U 1

Further possibilities for the transformation of data are

$$x'''_{ik} = f_k \, x_{ik}, \tag{2.1.37}$$

where a different weight $f_k > 0$ is assigned to each variable, and the transformation

$$x''''_{ik} = \frac{x_{ik} - a_k}{b_k - a_k} \tag{2.1.38}$$

where $\quad a_k = \min_i x_{ik}$

and $\quad b_k = \max_i x_{ik},$

in which the smallest element of each column becomes equal to 0 and the largest becomes equal to 1, corresponding to

$$\|x''''_{.k}\|_\infty = 1 \tag{2.1.39}$$

Transformations using

$$\|x_{.k}\|_1 = 1 \quad \text{or} \quad \|x_{i.}^T\|_1 = 1 \tag{2.1.40}$$

are also worth considering. Also used are transformations of the type (2.1.37) combined with (2.1.36), and non-linear transformations (see Wallace 1968).

For many cluster algorithms the most appropriate method of scaling is not known in advance, and a number of methods may be tried. Those for which the clusters are particularly easy to interpret are regarded as meaningful. A suitable normalisation of the variables conforming to (2.1.36), possibly followed by a suitable weighting (2.1.37), will usually be used. Normalisation of the objects is often not effective because the relationship (2.1.29) then applies.

In order to gain insight into the problems of scaling, we shall consider the artificial example shown in Fig. B1. Four points appear in three different simple transformations of the Cartesian coordinate system. One would, purely by intuition, form four clusters in the first case, where a single point is assigned to each; in the second case, one would form two clusters consisting of the two pairs (1,2) and (3,4); and in the third case, one would form two clusters consisting of the two pairs of points (1,3) and (2,4). Of course, in practical cases the situation is not so extreme.

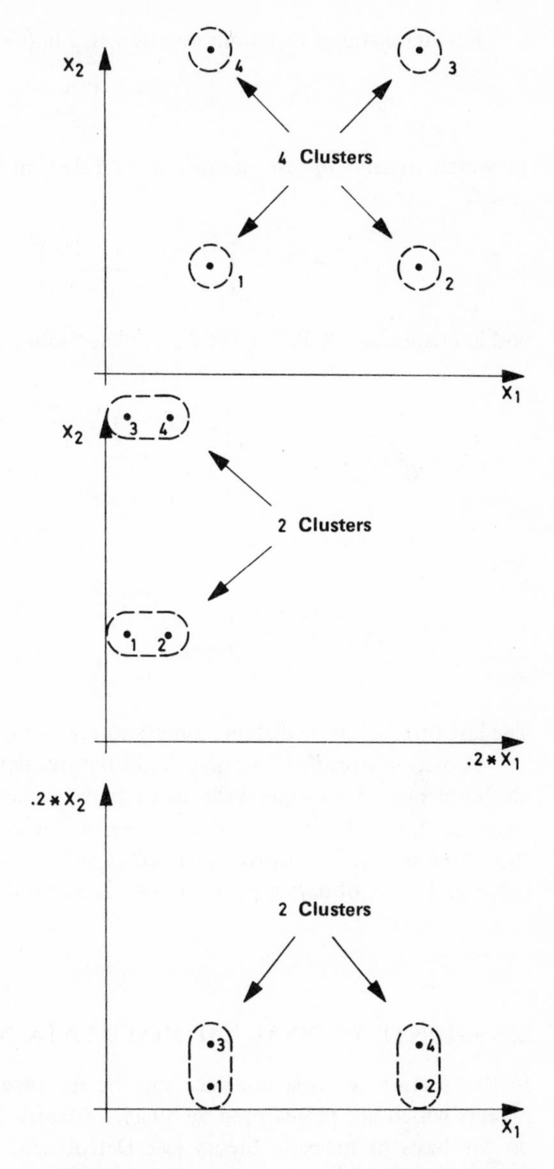

Fig. B 1

We shall not concern ourselves further with the question of effective scaling, since it cannot be dealt with in general terms. From Chapter 3 onwards we shall presume either that a decision about scaling has been made, or that, as an alternative, a variety of scaling methods will be used. At the same time, in Chapter 3, we discuss scale invariant cluster algorithms, which will certainly necessitate a great deal of calculation and none of which will be singled out as superior to the others.

Further distance functions are discussed in (Gower 1967):

$$d_G(x_{i.}^T, x_{j.}^T) = -\log_{10}\left(1 - \frac{1}{l}\sum_{k=1}^{l}\frac{|x_{ik} - x_{jk}|}{b_k - a_k}\right), \tag{2.1.41}$$

in which a_k and b_k are defined in (2.1.38); in (Sokal and Sneath 1963) for $X \geqslant 0$

$$d_Q(x_{i.}^T, x_{j.}^T) = \sqrt{\frac{1}{l}\sum_{k=1}^{l}\left(\frac{x_{ik} - x_{jk}}{x_{ik} + x_{jk}}\right)^2} \tag{2.1.42}$$

and in (Lance and Williams 1967c, 1968a, 1968b, 1971):

$$d_L(x_{i.}^T, x_{j.}^T) = \frac{\sum_{k=1}^{l}|x_{ik} - x_{jk}|}{\sum_{k=1}^{l}|x_{ik} + x_{jk}|} \tag{2.1.43}$$

and

$$d_M(x_{i.}^T, x_{j.}^T) = \sum_{k=1}^{l}\frac{|x_{ik} - x_{jk}|}{|x_{ik}| + |x_{jk}|}. \tag{2.1.44}$$

The last three of these distance functions are not metric.

Choice of distance functions is, as before, determined by the success of the cluster algorithm. In general the use of metric distance functions is to be preferred. Distance functions like d_1 and d_2^2, in which the contribution of each variable to the whole is purely additive, are used especially with mixed data, and whenever individual items of data are missing (see Lance and Williams 1967).

2.2 NOMINAL, ORDINAL AND MIXED DATA

In this section we first consider specific distance and similarity functions for objects which are represented by binary variables. Such functions can be derived on the basis of measure theory (see Dattola and Murray 1967, Deichsel 1972, Fritsche 1973, Green and Rao 1969, Majone and Sanday 1968, Restle 1961, Soergel 1967) or in an ad hoc manner (see Anderberg 1973, Cheetham and Hazel 1969, Green and Rao 1969, Harrison 1968, Joyce and Channon 1966, Reitema and Sagalyn 1967, Rogers and Tanimoto 1969, Sokal and Sneath 1963). Here we shall adopt the second method, which is shorter, though it does make it necessary to demonstrate explicitly the metric characteristics of the most important and most frequently applied distance and similarity functions.

From now on, until stated otherwise, the vectors x, y and z, with elements x_k, y_k and z_k (k=1, . . . ,l) such that each element is either 0 or 1, will denote rows of the binary data matrix (1.1).

The following abbreviations will be used to shorten the definitions of the similarity functions. For two binary vectors x and y, let

$$\alpha = \sum_{k=1}^{l} \min(x_k, y_k) \tag{2.2.1}$$

denote the number of corresponding positions in the vectors where a 1 appears in both x and y,

$$\beta = \sum_{k=1}^{l} x_k - \alpha \tag{2.2.2}$$

the number of positions with a 0 in x and a 1 in y,

$$\gamma = \sum_{k=1}^{l} y_k - \alpha \tag{2.2.3}$$

the number of positions with a 1 in x and a 0 in y, and

$$\delta = l - (\alpha + \beta + \gamma) \tag{2.2.4}$$

the number of positions with a 0 appearing in both x and y.

First we distinguish (see Sokal and Sneath 1973) between similarity functions which contain only α, i.e. matching of the 1s, and those which contain α and δ equally weighted, i.e. matching of the 1s and matching of the 0s are given equal value. We begin by setting out, on the left and right hand sides below, six pairs of corresponding similarity functions of this kind:

$$\frac{\alpha}{l} \qquad\qquad \frac{\alpha+\delta}{l} \tag{2.2.5}$$

$$\frac{\alpha}{\alpha+\beta+\gamma} \qquad\qquad \frac{\alpha+\delta}{\alpha+\beta+\gamma+\delta} \tag{2.2.6}$$

$$\frac{\alpha}{\alpha+2(\beta+\gamma)} \qquad\qquad \frac{\alpha+\delta}{\alpha+\delta+2(\beta+\gamma)} \tag{2.2.7}$$

$$\frac{2\alpha}{2\alpha+(\beta+\gamma)} \qquad\qquad \frac{2(\alpha+\delta)}{2(\alpha+\delta)+\beta+\gamma} \tag{2.2.8}$$

$$\frac{1}{2}\left(\frac{\alpha}{\alpha+\beta}+\frac{\alpha}{\alpha+\gamma}\right) \qquad\qquad \frac{1}{4}\left(\frac{\alpha}{\alpha+\beta}+\frac{\alpha}{\alpha+\gamma}+\frac{\delta}{\gamma+\delta}+\frac{\delta}{\beta+\delta}\right) \tag{2.2.9}$$

$$\frac{\alpha}{\sqrt{(\alpha+\beta)(\alpha+\gamma)}} \qquad \frac{\alpha\delta}{\sqrt{(\alpha+\beta)(\alpha+\gamma)(\delta+\beta)(\delta+\gamma)}}. \tag{2.2.10}$$

The similarity functions on the right hand side are defined so that they are symmetrical in α and δ. Since $s_0 = 1$ for all the similarity functions in (2.2.5)-(2.2.10), the functions $d = 1-s$ are distance functions and, as previously noted, if s is metric then d is also metric, and vice versa.

Let us consider the distance functions (2.2.5) and (2.2.6). For these it holds that

$$\frac{\alpha}{l} \leqq \frac{\alpha}{\alpha+\beta+\gamma} \leqq \frac{\alpha+\delta}{\alpha+\beta+\gamma+\delta} \tag{2.2.11}$$

The first part is evident from the fact that $l = \alpha + \beta + \gamma + \delta$, and the second follows from the relation, also used again later,

$$\frac{a}{c} \leqq \frac{a+b}{c+b} \quad \text{for } c \geqq a,\ b > 0, \tag{2.2.12}$$

putting $a = \alpha$, $b = \beta$, and $c = \alpha + \beta + \gamma$.

The quantity α/l gives the proportion of positions which have a 1 in both of the binary vectors x and y, taken over their total length; in $\alpha/(\alpha + \beta + \gamma)$ this proportion is calculated only for the positions which are not simultaneously 0 for both vectors; and $(\alpha + \delta)/l$ gives the proportion, taken over the total length, of positions which are simultaneously 1 or simultaneously 0 in both vectors. This last similarity function is metric (see Anderberg 1973), since $(\beta + \gamma)/l$ corresponds to the L_1 metric. It is not an appropriate one to use if the simultaneous appearance of 1 (matching relative to one variable) needs to be given greater weight than the simultaneous appearance of 0 in the same position, which is certainly the case if zeros predominate in the binary data matrix. In this case the first coefficient α/l always takes on very small values. The so-called "Tanimoto coefficient" $\alpha/(\alpha + \beta + \gamma)$ appears to be the ideal compromise, particularly since it is a metric similarity function, in contrast to (2.2.8)-(2.2.10) and to the similarity functions which follow later. Before formally demonstrating the metric properties, we give some illustrations to provide further foundations for the previous statements. If we have

$$x = (1,1,0,1,1,0,1)^T$$
$$y = (1,0,1,0,0,0,1)^T$$
$$z = (1,0,1,1,1,0,0)^T$$

then we obtain

	$\frac{\alpha}{l}$	$\frac{\alpha}{\alpha+\beta+\gamma}$	$\frac{\alpha+\delta}{l}$
$s(x, y)$	$^2/_7$	$^1/_3$	$^3/_7$
$s(y, z)$	$^2/_7$	$^2/_5$	$^4/_7$
$s(x, z)$	$^3/_7$	$^1/_2$	$^4/_7$

It is clear that, for the distance function

$$d_T(x, y) = 1 - \frac{\alpha}{\alpha+\beta+\gamma} = \frac{\beta+\gamma}{\alpha+\beta+\gamma} \tag{2.2.13}$$

the properties (2.1.1)-(2.1.3) hold when $d_0 = 0$. We now demonstrate the metric properties (2.1.4) and (2.1.5). From $d_T(x,y) = 0$ it follows that

$$\beta+\gamma = \left(\sum_{k=1}^{l} x_k - \sum_{k=1}^{l} \min(x_k, y_k)\right) + \left(\sum_{k=1}^{l} y_k - \min(x_k, y_k)\right) = 0.$$

Since the components x_k and y_k are non-negative, each term of the sum is non-negative and, moreover, it must certainly hold that

$$\sum_{k=1}^{l} x_k - \sum_{k=1}^{l} \min(x_k, y_k) = 0$$

and

$$\sum_{k=1}^{l} y_k - \sum_{k=1}^{l} \min(x_k, y_k) = 0$$

This, however, is only possible if $x_k = y_k$, $(k = 1, \ldots, l)$, that is, if $x = y$.

In order to prove the triangular inequality, we require in addition to (2.2.12) the self-evident inequalities

$$\sum_{k=1}^{l} \min(x_k, y_k) \geqq \sum_{k=1}^{l} \min(x_k, y_k, z_k) \tag{2.2.14}$$

and

$$\sum_{k=1}^{l} z_k - \sum_{k=1}^{l} \min(y_k, z_k) - \sum_{k=1}^{l} \min(x_k, z_k) + \sum_{k=1}^{l} \min(x_k, y_k, z_k) \geqq 0, \tag{2.2.15}$$

which are certainly valid for any values of x_k, y_k, $z_k \geqslant 0$.

We can now demonstrate the triangular inequality:

$$d_T(x, y) + d_T(y, z)$$

$$= \frac{\sum x_k + \sum y_k - 2\sum \min(x_k, y_k)}{\sum x_k + \sum y_k - \sum \min(x_k, y_k)} + \frac{\sum y_k + \sum z_k - 2\sum \min(y_k, z_k)}{\sum y_k + \sum z_k - \sum \min(y_k, z_k)}$$

$$= \frac{\begin{array}{l}\sum x_k + \sum \min(y_k, z_k) - \sum \min(x_k, y_k) - \sum \min(x_k, y_k, z_k) \\ +[\sum y_k - \sum \min(y_k, z_k) - \sum \min(x_k, y_k) + \sum \min(x_k, y_k, z_k)]\end{array}}{\begin{array}{l}\sum x_k + \sum \min(y_k, z_k) \qquad\qquad - \sum \min(x_k, y_k, z_k) \\ +[\sum y_k - \sum \min(y_k, z_k) - \sum \min(x_k, y_k) + \sum \min(x_k, y_k, z_k)]\end{array}}$$

$$+ \frac{\begin{array}{l}\sum z_k + \sum \min(x_k, y_k) - \sum \min(y_k, z_k) - \sum \min(x_k, y_k, z_k) \\ +[\sum y_k - \sum \min(x_k, y_k) - \sum \min(y_k, z_k) + \sum \min(x_k, y_k, z_k)]\end{array}}{\begin{array}{l}\sum z_k + \sum \min(x_k, y_k) \qquad\qquad - \sum \min(x_k, y_k, z_k) \\ +[\sum y_k - \sum \min(x_k, y_k) - \sum \min(y_k, z_k) + \sum \min(x_k, y_k, z_k)]\end{array}}$$

$$\geq \frac{\sum x_k + \sum \min(y_k, z_k) - \sum \min(x_k, y_k) - \sum \min(x_k, y_k, z_k)}{\sum x_k + \sum \min(y_k, z_k) \qquad\qquad - \sum \min(x_k, y_k, z_k)}$$

$$+ \frac{\sum z_k + \sum \min(x_k, y_k) - \sum \min(y_k, z_k) - \sum \min(x_k, y_k, z_k)}{\sum z_k + \sum \min(x_k, y_k) \qquad\qquad - \sum \min(x_k, y_k, z_k)}$$

$$\geq \frac{\sum x_k + \sum \min(y_k, z_k) - \sum \min(x_k, y_k) - \sum \min(x_k, y_k, z_k)}{\sum x_k + \sum z_k \qquad\qquad - \sum \min(x_k, z_k)}$$

$$+ \frac{\sum z_k + \sum \min(x_k, y_k) - \sum \min(y_k, z_k) - \sum \min(x_k, y_k, z_k)}{\sum z_k + \sum x_k \qquad\qquad - \sum \min(x_k, z_k)}$$

$$= \frac{\sum x_k + \sum z_k - 2\sum \min(x_k, y_k, z_k)}{\sum z_k + \sum x_k - \sum \min(x_k, y_k)}$$

$$\geq \frac{\sum x_k + \sum z_k - 2\sum \min(x_k, z_k)}{\sum x_k + \sum z_k - \sum \min(x_k, z_k)}$$

$$= d_T(x, z).$$

We shall now return to the similarity functions (2.2.7) to (2.2.10), which have not yet been discussed. The distance functions

$$d_U(x, y) = \frac{2(\beta+\gamma)}{\alpha+2(\beta+\gamma)} \tag{2.2.16}$$

and

$$d_V(x, y) = \frac{\beta+\gamma}{2\alpha+(\beta+\gamma)} \tag{2.2.17}$$

which are analogous to the right-hand expressions in (2.2.7) and (2.2.8), differ in that, in the first case, the number of positions with a 0 and 1 and the number with a 1 and a 0 the two vectors have double weighting, while in the second case, the number of positions with matching 1s is double weighted. Since, further,

$$d_U(x, y) = \frac{2\, d_T(x, y)}{1 + d_T(x, y)}, \tag{2.2.18}$$

then d_U is also metric along with d_T (see Majone and Sanday 1968), whereas no similar transformation can be found for d_V. The relation

$$d_V(x, y) = \frac{2(1 - d_T(x, y))}{2 - d_T(x, y)}, \tag{2.2.19}$$

certainly holds, but the retention of the metric properties cannot, however, yet be proved for this transformation. The same is true for the righthand sides of (2.2.7) (see Anderberg 1973) and (2.2.8).

The interpretation of the similarity functions (2.2.9), based on the meaning of α, β, γ and δ is evident. The formulae (2.2.10) correspond to correlation measures (see Anderberg 1973).

Other non-metric similarity functions with a range of values $[-1,1]$ are

$$\frac{\alpha+\delta-(\beta+\gamma)}{l}, \tag{2.2.20}$$

$$\frac{\alpha\delta-\beta\gamma}{\alpha\delta+\beta\gamma}, \tag{2.2.21}$$

$$\frac{\alpha\delta-\beta\gamma}{\sqrt{(\alpha+\beta)(\alpha+\gamma)(\beta+\delta)(\gamma+\delta)}}. \tag{2.2.22}$$

Exceptional cases are to be found in (Cheetham and Hazel 1969) and in (Reitsma and Sagalyn 1967). We notice in addition that

$$d(x, y) = \sqrt{\frac{\beta+\gamma}{l}} \tag{2.2.23}$$

corresponds to the formal Euclidean distance multiplied by $\sqrt{\frac{1}{l}}$. Furthermore, as already mentioned,

$$d(x, y) = \frac{1}{l}(\beta + \gamma)$$

corresponds to the L_1 metric and, moreover, to d_p^p for all $p \geqslant 1$.

In general we agree with the views frequently expressed in the literature that in practical applications d_T is of particular value, both because of its emphasis on matching (of 1s) and because of its metric properties.

Instead of binary variables, let us now consider a nominal variable with more than two states. Then

$$1 \leqq x_k \leqq j_k \quad (k = 1, \dots, l), \tag{2.2.24}$$

where x_k and j_k are integer, holds for the components $x_1, \dots, x_l$ of the data matrix (1.1), which is now nominal.

We now map the vector x of length l into a binary vector $\tilde{x}$ of length $l' = \sum_{k=1}^{l} j_k$ by the transformation

$$x \longrightarrow \tilde{x} = (\underbrace{\overset{x_1}{0, \dots, 1, \dots, 0}}_{j_1}; \dots; \underbrace{\overset{x_l}{0, \dots, 1, \dots, 0}}_{j_l})^T \tag{2.2.25}$$

(see Talkington 1967). In $\tilde{x}$ there are l binary vectors of length j_k which now have a 1 in the x_k -th place and 0s elsewhere. For example, we get the vector $x = (0,1,0;0,0,1,0;1,0)^T$ from the vector $x = (2,3,1)^T$ for $l = 3, j_1 = 3, j_2 = 4, j_3 = 2$. In this way nominal vectors are mapped onto binary vectors in one to one correspondence, and the distance and similarity functions for binary vectors which were described earlier are applicable (see Green and Tull 1970, Sokal and Sneath 1963).

If the j_k values are greatly different, so that the corresponding weights for each variable might be adequate for the purposes of matching, (Harrison 1968) recommends the similarity function

$$s_H(x, y) = \frac{\sum_{k=1}^{l} \ln j_k \cdot \tau(x_k, y_k)}{\sum_{k=1}^{l} \ln j_k} \tag{2.2.26}$$

where $\tau(x_k, y_k) = \begin{cases} 0 & x_k \neq y_k \\ 1 & x_k = y_k \end{cases}$

with the weights so chosen that they are proportional to the information in the kth variable. Only the matching is counted in s_H. $s_H(x,y) = \alpha/l$ (2.2.5) results again for $j_K = 2$ $(k = 1, \ldots, l)$.

If the data matrix contains ordinal variables only, a range

$$0 \leqq x_k,\ y_k \leqq r_k, \tag{2.2.27}$$

analogous with (2.2.24) is valid for the components of the two row vectors $x^T = (x_1, \ldots, x_l)$ and $y^T = (y_1, \ldots, y_l)$. The zero on the left side can be obtained, without loss of generality, by means of a translation. The distinction between this and (2.2.24) is that, in the case of ordinal values, one of the statements $x_k \geqslant y_k$ or $x_k < y_k$ must be true and hence the expression $\min(x_k, y_k)$ is particularly meaningful. If one defines

$$d_O(x,y) = \frac{\sum_{k=1}^{l} x_k + \sum_{k=1}^{l} y_k - 2\sum_{k=1}^{l} \min(x_k, y_k)}{\sum_{k=1}^{l} x_k + \sum_{k=1}^{l} y_k - \sum_{k=1}^{l} \min(x_k, y_k)}, \tag{2.2.28}$$

as a distance function (see Soergel 1967), this definition corresponds to a previous distance function derived from the similarity function $\alpha/(\alpha + \beta + \gamma)$ with binary data by the use of the transfomations $d_T = 1 - s$. The function d_O is metric, as one recognises in the proof of the metric characteristics of d_T. Furthermore,

$$0 \leqq d_O(x,y) \leqq 1. \tag{2.2.29}$$

holds. For example, given the vectors $x = (2,4,5,1)^T$, $y = (3,1,1,5)^T$ and $z = (2,3,4,2)^T$, then $d_O(x,y) = 12/17$, $d_O(x,z) = 3/13$ and $d_O(y,z) = 3/5$.

If as usual $1 \leqslant x_k \leqslant j_k$ (x_k, j_k integer, $k = 1, \ldots, l$), a mapping of the ordinal vector x on a binary vector $\tilde{x}$ of length $l' = \sum_{k=1}^{l} j_k$ is possible, namely

$$x \longrightarrow \tilde{x} = (\underbrace{\overset{x_1}{1, 1, \ldots, 1}, 0, \ldots, 0}_{j_1}; \ldots; \underbrace{\overset{x_l}{1, 1, \ldots, 1}, 0, \ldots, 0}_{j_l})^T, \tag{2.2.30}$$

(see Sokal and Sneath 1963) and we have

$$d_T(\tilde{x}, \tilde{y}) = d_O(x,y). \tag{2.2.31}$$

For example, for $j_k = 5$ $(k = 1, \ldots, 4)$ the binary vectors

$$\widetilde{x} = (1,1,0,0,0;\ 1,1,1,1,0;\ 1,1,1,1,1;\ 1,0,0,0,0)^T$$

$$\widetilde{y} = (1,1,1,0,0;\ 1,0,0,0,0;\ 1,0,0,0,0;\ 1,1,1,1,1)^T,$$

belong in this way to the vectors x and y above, and this yields

$$d_T(\widetilde{x}, \widetilde{y}) = d_O(x, y) = \frac{12}{17}.$$

The semi-ordering is preserved by the mapping (2.2.30) because, if $x \geqslant y$ holds element by element, then $\widetilde{x}$ is lexicographically greater than $\widetilde{y}$.

It is possible to generalise others of the similarity functions for binary data that were given earlier. We are not, however, going into them further, since we consider d_O, similar to the earliest d_T, to be the metric for ordinal data which is most important in practical cases; it has been used very successfully, for example, in the classification of symptoms (see Langenmayr and Späth 1977).

Now we wish to discuss the case in which the data matrix (1.1) contains variables of nominal, ordinal and metric types, mixed. We can permute the variables in such a way that blocks of the different types occur. Two row vectors of the data matrix can then be described as

$$x = \begin{pmatrix} x^{(N)} \\ x^{(O)} \\ x^{(M)} \end{pmatrix} \quad \text{and} \quad y = \begin{pmatrix} y^{(N)} \\ y^{(O)} \\ y^{(M)} \end{pmatrix} \tag{2.2.32}$$

in which N refers to nominal (and therefore, in special cases, binary) variables, O to ordinal, and M to metric variables. Variables of any one type could be missing.

A first possibility in defining a distance function d_A, which we recommend because it is additive, is then

$$\begin{aligned} d_A(x, y) = {} & p_1\, d^{(N)}(x^{(N)}, y^{(N)}) \\ & + p_2\, d^{(O)}(x^{(O)}, y^{(O)}) \\ & + p_3\, d^{(M)}(x^{(M)}, y^{(M)}), \end{aligned} \tag{2.2.33}$$

(see Green and Carmone 1970), in which p_1, p_2 and p_3 are suitably normalised positive weights and $d^{(N)}$, $d^{(O)}$ and $d^{(M)}$ are distance functions for nominal, ordinal and metric data respectively. If these are metric, then d_A is likewise a metric distance function. The same holds for similarly constructed similarity functions.

A second possibility (see Anderberg 1973, Green and Carmone 1970, Green and Tull 1970, Rubin 1967, Sokal and Sneath 1963) which may, however, lead to loss of information, is to convert the vectors x and y from (2.2.32) into binary vectors and use the defined distance functions for this instead. We have already seen in (2.2.25) how one transforms a nominal variable with more than two states into a binary variable. With ordinal data one can undertake a mapping on binary vectors according to (2.2.30), or – with loss of information – one can map an ordinal onto a binary variable, in which 0 is assigned to all values above the median and 1 is assigned to all values which are equal to or greater than the median. This mapping is irreversible. Metric variables can be treated in two ways. The first is similar to that for ordinal variables, except that the mean value is used rather than the median. The second is to map them onto ordinal variables on the basis of the appropriate mean value limits. The choice of transformations depends also on what mixture of types of variable appears in the data matrix; the ratios of the mixture may also play a part, or give rise to weightings analogously to (2.2.26). In (Anderberg 1973) many different kinds of data conversions are presented and worked through.

A third possibility is discussed in (Parks 1969). All variables are so transformed that their range of values is the interval (0,1) and then the Euclidean metric is used, modified by a factor

$$d_2(\tilde{x}, \tilde{y}) = \sqrt{\frac{\sum_{k=1}^{l} (\tilde{x}_k - \tilde{y}_k)^2}{l}} \qquad (2.2.34)$$

For binary data there is nothing to transform. Nominal data could, as described earlier, be converted into binary. An ordinal variable which assumes the values $1, \ldots, j$ is, for example, transformed in such a way that it assumes the values $0, 1/(j-1), 2/(j-1), \ldots, 1$. For metric data the transformation (2.1.38) is applied.

In dealing with mixed data, the use of mean values must be avoided as soon as nominal or ordinal variables appear. Therefore, in these cases, clustering methods will be used that depend on calculated distances and similarities of the objects, and do not involve the calculation of the mean from the (mixed) data matrix (1.1).

A further problem is encountered if variable values are missing from the data matrix for some of the objects, whether these values are of one type only or are mixed. In this case (Rubin 1967) recommends that, in calculating a distance matrix, no account should be taken of those objects, or of variables for which only one object has been measured. Suitable weights are useful (see Bock 1974).

We next come to another, quite different, kind of distance function (see Green and Tull 1970, Green and Rao 1969), which is not calculated from a data matrix (1.1) for the objects, but where several variables are implicitly taken into account but not quantified. All possible pairs of objects could, for example, be placed in order according to their degree of similarity. If one assigns the number 1 to the most similar pair, the number 2 to the second most similar pair, and so on until the number $m(m-1)2$ is assigned to the last pair, the expression

$$d_R(x_{i.}^T, x_{j.}^T) = k, \quad k \in \left\{1, \ldots, \frac{m(m-1)}{2}\right\} \tag{2.2.35}$$

is therefore a distance function.

We have already pointed out in the introduction that for these types of rank order distances and for all distance matrices that are made up of nominal, ordinal and mixed data rather than metric, the techniques of non-metric and metric multidimensional scaling can be employed to give an approximately metric data matrix for the objects which can then be used in the cluster analysis.

Furthermore one can, at the cost in some circumstances of significant loss of information, produce from a distance matrix a matrix G containing only 0s and 1s by means of

$$d(x_{i.}^T, y_{j.}^T) \longrightarrow g_{ij} = \begin{cases} 1 & \text{when } d(x_{i.}^T, x_{j.}^T) \geqq \tau \\ 0 & \text{when } d(x_{i.}^T, x_{j.}^T) < \tau \end{cases} \tag{2.2.36}$$

and interpret this as an incidence matrix of a graph with m edges; i.e. the edges i and j are linked if $g_{ij} = 1$ and not linked for $g_{ij} = 0$. In section 5.3 we refer briefly to the clustering techniques used in this case.

Finally in this section, we shall take up the definition of distance and similarity functions for clusters of objects whose values are frequently referred to for interpretation. Let the objects be serially numbered from 1 up to m and let a cluster with the serial number p be defined here as a non-empty set of indices $C_p \subset \{1, \ldots, m\}$. It consists of m_p objects $x_{i.}$ with $i \in C_p$. Since C_p can consist of one element, we have additionally defined distances from clusters to objects.

Frequently cluster distances are defined implicitly by algorithms. We postulate herewith a few explicit values (see Bock 1970, Deichsel 1970, Fritsche 1973).

If a metric data matrix is given, the mean

$$\bar{x}_{p.}^T = \frac{1}{m_p} \sum_{i \in C_p} x_{i.}^T \tag{2.2.37}$$

can be used to represent the objects contained in the cluster C_p. Usual distance measures are then

$$d_M\,(C_p, C_q) = \|\bar{x}_{p.}^T - \bar{x}_{q.}^T\|_2, \tag{2.2.38}$$

$$d_{M'}\,(C_p, C_q) = \sqrt{\frac{m_p\, m_q}{m_p + m_q}}\, \|\bar{x}_{p.}^T - \bar{x}_{q.}^T\|_2, \tag{2.2.39}$$

$$d_S\,(C_p, C_q) = \sqrt{(\bar{x}_{p.} - \bar{x}_{q.})\, S^{-1}\, (\bar{x}_{p.} - \bar{x}_{q.})^T}, \tag{2.2.40}$$

in which S is defined in (2.1.26). d_M is the Euclidean distance of the cluster centroids, $d_M{}'$ the same but multiplied by a function of the numbers of elements, the exact significance of which will become clear later, and d_S a generalised Mahalanobis distance.

For non-metric distance functions

$$d_Z\,(C_p, C_q) = \min_{\substack{i \in C_p \\ j \in C_q}} d_Z\,(x_{i.}^T, x_{j.}^T) \tag{2.2.41}$$

is sometimes possible and analogously for non-metric similarity functions

$$s_Z\,(C_p, C_q) = \max_{\substack{i \in C_p \\ j \in C_q}} s_Z\,(x_{i.}^T, x_{j.}^T), \tag{2.2.42}$$

in which the index Z should refer to arbitrary distance and similarity functions on the righthand side.

Further possibilities are

$$s_Z\,(C_p, C_q) = \sum_{i \in C_p} \sum_{j \in C_q} s_Z\,(x_{i.}^T, x_{j.}^T) \tag{2.2.43}$$

respectively weighted

$$s_Z\,(C_p, C_q) = \frac{1}{m_p\, m_q} \sum_{i \in C_p} \sum_{j \in C_q} s_Z\,(x_{i.}^T, x_{j.}^T). \tag{2.2.44}$$

In this chapter we have introduced all the basic concepts; in the chapter which follows we shall concern ourselves with the cluster algorithms themselves. As already noted, we shall always assume that there is available for the objects either an appropriate scaled metric data matrix (1.1), or a distance matrix.

CHAPTER 3

Partitioning Cluster Algorithms

3.1 HEURISTIC METHODS

At the end of the last chapter we introduced the representation of a cluster with serial number j as a set of indices $C_j \subset \{1, \ldots, m\}$, where the objects are consecutively numbered from 1 to m.

By a partition of length n of m objects will be meant a collection of such non-empty sets C_j with

$$C_1 \cup C_2 \cup \ldots \cup C_n = \{1, \ldots, m\} \tag{3.1.1}$$

$$C_j \cap C_k = \phi \quad (j \neq k), (\phi \text{ the empty set}) \tag{3.1.2}$$

We represent a partition of length n by an assignment vector p of length m; if the ith element, $p_i = j$ it means that the ith object belongs to the jth cluster; $p_i = 0$ means that the ith object does not yet belong to a cluster or that in fact it belongs to no cluster whatsoever. Since it was postulated that the above sets of indices should not be empty, let us assume correspondingly that, for the assignment vector p, for each $j = 1, \ldots, n$, at least one $i \in \{1, \ldots, m\}$ with $p_i = j$ exists. Values of p_i other than $p_i = 0$ or for which $1 \leqslant p_i \leqslant n$ are not admissible.

In general, in this chapter we shall try to find a partitioning, with a given n, which is optimal with respect to a given objective function. However, in this first section, we are proceeding heuristically, i.e. without an objective function, and must allow

$$C_1 \cup C_2 \cup \ldots \cup C_n \subset \{1, \ldots, m\} \tag{3.1.3}$$

instead of (3.1.1), or begin without an assumed value of n.

There are an arbitrarily large number of heuristic procedures (see Anderberg 1973). All have in common the way in which they are oriented towards visual, geometric representation of cluster formations of point objects distributed in a plane. This is very reasonable, since visual classification of points in a plane cannot be improved upon by algorithms. In fact cluster algorithms are superior to visual classification only for more than two dimensions.

Of the heuristic algorithms, the simplest is based on considering each object just once and immediately allocating it to a cluster. This algorithm, applied to a metric data matrix X(M,L), is given in the subroutine LEADER (Fig. U2). RHO signifies a threshold value in the allocation; P, as always in what follows, stands for the assignment vector, already described, and NMAX stands for the maximum number of clusters to be formed. If RHO is so small that NMAX

```
      SUBROUTINE LEADER (M,L,X,NMAX,RHO,P)
C
C     AD HOC CONSTRUCTION OF CLUSTERS WITHOUT SPECIAL OPTIMAL
C     CHARACTERISTICS USING THE FOLLOWING ALGORITHM:
C
C     1.    N = 1, I = 1.
C     2.    X(I) IS ALLOCATED TO THE N-TH CLUSTER.
C     3.    I = I + 1, STOP IF I > M.
C     4.    ASSIGN X(I) TO THE FIRST OF THE PREVIOUSLY GENERATED
C           CLUSTERS FOR WHICH THE FIRST (LEADING) ELEMENT IS AT
C           A DISTANCE LESS THAN RHO.
C     5.    IF THERE IS NO SUCH CLUSTER AND N < NMAX, SET
C           N = N + 1 AND RETURN TO STEP 2, OTHERWISE LEAVE X(I)
C           UNALLOCATED AND RETURN TO STEP 3.
C
C     P(I) CONTAINS THE NUMBER OF THE CLUSTER TO WHICH X(I) HAS
C     BEEN ASSIGNED.   P(I) = 0 MEANS NO ASSIGNMENT.
C


C     DIMENSION X(M,L),    P(M),    F(NMAX)
      DIMENSION X(100,12),P(100),F(10)
      INTEGER    P,F,R
      N=1
      F(1)=1
      DO 3 I=1,M
           P(I)=0
           DO 2 J=1,N
                R=F(J)
                D=0.
                DO 1 K=1,L
                     T=X(R,K)-X(I,K)
                     D=D+T*T
    1           CONTINUE
                IF(SQRT(D).GT.RHO) GOTO 2
                P(I)=J
                GOTO 3
    2      CONTINUE
           IF(N.EQ.NMAX) GOTO 3
           N=N+1
           P(I)=N
           F(N)=I
    3 CONTINUE
      RETURN
      END
```

Fig. U 2

clusters are insufficient to allocate all the points, the remaining points are not assigned, and therefore the result depends on the sequence in which the objects are taken. If RHO is too large, fewer than NMAX clusters are formed, which can be recognised from the values of the assignment vector P produced.

In the main program H1, for each NMAX (the number of clusters to be formed) from N1, N1+1, . . . ,N2 the LEADER subroutine is repeatedly called, using RHO=J*DELTA (J=1, . . . ,JMAX), JMAX being determined as the first J for which all objects are assigned to clusters ($p_i \neq 0$ for $i = 1, \ldots, m$).

```
C
C     TEST PROGRAM FOR THE LEADER ALGORITHM
C
      DIMENSION X(100,12),P(100)
      INTEGER P
      KI=5
      KO=6
    1 READ(KI,2) M,L,N1,N2,DELTA
    2 FORMAT(4I5,F5.0)
      IF(M.LE.0.OR.M.GT.100.OR.L.LT.1.OR.L.GT.12
     *    .OR.N1.LE.1.OR.N2.GT.10) STOP
      IF(DELTA.LE.0.) DELTA=1.
      WRITE(KO,3) M,L,DELTA
    3 FORMAT('1',' M=',I3,'  L=',I2,'  DELTA=',F6.1)
      WRITE(KO,4)
    4 FORMAT('0')
      DO 5 K=1,L
          READ(KI,6)  (X(I,K),I=1,M)
          WRITE(KO,7) (X(I,K),I=1,M)
    5 CONTINUE
      WRITE(KO,4)
      WRITE(KO,11) (I,I=1,M)
      WRITE(KO,4)
    6 FORMAT(16F5.0)
    7 FORMAT(3X,10F6.1)
      N=N1
      RHO=0.
    8 RHO=RHO+DELTA
      CALL LEADER (M,L,X,N,RHO,P)
      WRITE(KO,9) N,RHO,(P(I),I=1,M)
    9 FORMAT(1X,I3,F6.1,23I3/(15X,23I3))
      DO 10 I=1,M
          IF(P(I).EQ.0) GOTO 8
   10 CONTINUE
   11 FORMAT(10X,23I3)
      RHO=0.
      N=N+1
      IF(N.LE.N2) GOTO 8
      GOTO 1
      END
```

Fig. H 1

The results for the case L = 1 and M = 15 for two different sequences of the $x_{i.}$, are set out in tables E1.1 and 1.2. NMAX, RHO and (P(I), I= 1,M) are displayed as in the output specification in H1. The results differ significantly because of the difference in ordering of the sequences, but both bear out the earlier remarks about the opposing influences of RHO and NMAX.

Fig. E 1.1

```
M= 15   L= 1   DELTA=     0.3

      0.0    0.1    0.2    0.5    2.0    2.1    2.5    3.0    3.2    4.0
      5.0    6.0    8.0   10.0   11.0
```

		1	2	3	4	5	6	7	8	9	10	11	12	13	14	15
3	0.3	1	1	1	2	3	3	0	0	0	0	0	0	0	0	0
3	0.6	1	1	1	1	2	2	2	3	3	0	0	0	0	0	0
3	0.9	1	1	1	1	2	2	2	3	3	0	0	0	0	0	0
3	1.2	1	1	1	1	2	2	2	2	2	3	3	0	0	0	0
3	1.5	1	1	1	1	2	2	2	2	2	3	3	0	0	0	0
3	1.8	1	1	1	1	2	2	2	2	2	3	3	0	0	0	0
3	2.1	1	1	1	1	1	2	2	2	2	2	3	3	0	0	0
3	2.4	1	1	1	1	1	1	2	2	2	2	3	3	0	0	0
3	2.7	1	1	1	1	1	1	1	2	2	2	2	3	3	0	0
3	3.0	1	1	1	1	1	1	1	2	2	2	2	3	3	0	0
3	3.3	1	1	1	1	1	1	1	1	1	2	2	2	3	3	3
4	0.3	1	1	1	2	3	3	4	0	0	0	0	0	0	0	0
4	0.6	1	1	1	1	2	2	2	3	3	4	0	0	0	0	0
4	0.9	1	1	1	1	2	2	2	3	3	4	0	0	0	0	0
4	1.2	1	1	1	1	2	2	2	2	2	3	3	4	0	0	0
4	1.5	1	1	1	1	2	2	2	2	2	3	3	4	0	0	0
4	1.8	1	1	1	1	2	2	2	2	2	3	3	4	0	0	0
4	2.1	1	1	1	1	1	2	2	2	2	2	3	3	4	4	0
4	2.4	1	1	1	1	1	1	2	2	2	2	3	3	4	4	0
4	2.7	1	1	1	1	1	1	1	2	2	2	2	3	3	4	4
5	0.3	1	1	1	2	3	3	4	5	5	0	0	0	0	0	0
5	0.6	1	1	1	1	2	2	2	3	3	4	5	0	0	0	0
5	0.9	1	1	1	1	2	2	2	3	3	4	5	0	0	0	0
5	1.2	1	1	1	1	2	2	2	2	2	3	3	4	5	0	0
5	1.5	1	1	1	1	2	2	2	2	2	3	3	4	5	0	0
5	1.8	1	1	1	1	2	2	2	2	2	3	3	4	5	0	0
5	2.1	1	1	1	1	1	2	2	2	2	2	3	3	4	4	5
6	0.3	1	1	1	2	3	3	4	5	5	6	0	0	0	0	0
6	0.6	1	1	1	1	2	2	2	3	3	4	5	6	0	0	0
6	0.9	1	1	1	1	2	2	2	3	3	4	5	6	0	0	0
6	1.2	1	1	1	1	2	2	2	2	2	3	3	4	5	6	6
7	0.3	1	1	1	2	3	3	4	5	5	6	7	0	0	0	0
7	0.6	1	1	1	1	2	2	2	3	3	4	5	6	7	0	0
7	0.9	1	1	1	1	2	2	2	3	3	4	5	6	7	0	0
7	1.2	1	1	1	1	2	2	2	2	2	3	3	4	5	6	6
8	0.3	1	1	1	2	3	3	4	5	5	6	7	8	0	0	0
8	0.6	1	1	1	1	2	2	2	3	3	4	5	6	7	8	0
8	0.9	1	1	1	1	2	2	2	3	3	4	5	6	7	8	0
8	1.2	1	1	1	1	2	2	2	2	2	3	3	4	5	6	6
9	0.3	1	1	1	2	3	3	4	5	5	6	7	8	9	0	0
9	0.6	1	1	1	1	2	2	2	3	3	4	5	6	7	8	9
10	0.3	1	1	1	2	3	3	4	5	5	6	7	8	9	10	0
10	0.6	1	1	1	1	2	2	2	3	3	4	5	6	7	8	9

```
M= 15   L= 1   DELTA=    0.3

   11.0    2.5    8.0    3.2    5.0    0.1   10.0    2.1   3.0    0.5
    2.0    6.0    0.0    0.2    4.0
```

		1	2	3	4	5	6	7	8	9	10	11	12	13	14	15
3	0.3	1	2	3	0	0	0	0	0	0	0	0	0	0	0	0
3	0.6	1	2	3	0	0	0	0	2	2	0	2	0	0	0	0
3	0.9	1	2	3	2	0	0	0	2	2	0	2	0	0	0	0
3	1.2	1	2	3	2	0	0	1	2	2	0	2	0	0	0	0
3	1.5	1	2	3	2	0	0	1	2	2	0	2	0	0	0	0
3	1.8	1	2	3	2	0	0	1	2	2	0	2	0	0	0	2
3	2.1	1	2	3	2	0	0	1	2	2	2	2	3	0	0	2
3	2.4	1	2	3	2	0	0	1	2	2	2	2	3	0	2	2
3	2.7	1	2	3	2	2	2	1	2	2	2	2	3	2	2	2
4	0.3	1	2	3	4	0	0	0	0	4	0	0	0	0	0	0
4	0.6	1	2	3	4	0	0	0	2	2	0	2	0	0	0	0
4	0.9	1	2	3	2	4	0	0	2	2	0	2	0	0	0	0
4	1.2	1	2	3	2	4	0	1	2	2	0	2	4	0	0	4
4	1.5	1	2	3	2	4	0	1	2	2	0	2	4	0	0	4
4	1.8	1	2	3	2	4	0	1	2	2	0	2	4	0	0	2
4	2.1	1	2	3	2	4	0	1	2	2	2	2	3	0	0	2
4	2.4	1	2	3	2	4	0	1	2	2	2	2	3	0	2	2
4	2.7	1	2	3	2	2	2	1	2	2	2	2	3	2	2	2
5	0.3	1	2	3	4	5	0	0	0	4	0	0	0	0	0	0
5	0.6	1	2	3	4	5	0	0	2	2	0	2	0	0	0	0
5	0.9	1	2	3	2	4	5	0	2	2	5	2	0	5	5	0
5	1.2	1	2	3	2	4	5	1	2	2	5	2	4	5	5	4
6	0.3	1	2	3	4	5	6	0	0	4	0	0	0	6	6	0
6	0.6	1	2	3	4	5	6	0	2	2	6	2	0	6	6	0
6	0.9	1	2	3	2	4	5	6	2	2	5	2	0	5	5	0
6	1.2	1	2	3	2	4	5	1	2	2	5	2	4	5	5	4
7	0.3	1	2	3	4	5	6	7	0	4	0	0	0	6	6	0
7	0.6	1	2	3	4	5	6	7	2	2	6	2	0	6	6	0
7	0.9	1	2	3	2	4	5	6	2	2	5	2	7	5	5	0
7	1.2	1	2	3	2	4	5	1	2	2	5	2	4	5	5	4
8	0.3	1	2	3	4	5	6	7	8	4	0	8	0	6	6	0
8	0.6	1	2	3	4	5	6	7	2	2	6	2	8	6	6	0
8	0.9	1	2	3	2	4	5	6	2	2	5	2	7	5	5	8
9	0.3	1	2	3	4	5	6	7	8	4	9	8	0	6	6	0
9	0.6	1	2	3	4	5	6	7	2	2	6	2	8	6	6	9
10	0.3	1	2	3	4	5	6	7	8	4	9	8	10	6	6	0
10	0.6	1	2	3	4	5	6	7	2	2	6	2	8	6	6	9

Fig. E 1.2

The second example, which we shall often use, takes as its data the Cartesian coordinates (L = 2) of M = 22 German towns, derived from a map. The names of the towns are shown in Fig. T1 and their coordinates in Fig. E1.3.

Table T1

1 Aachen	9 Karlsruhe	17 Regensburg
2 Augsburg	10 Kassel	18 Saarbrücken
3 Braunschweig	11 Kiel	19 Würzburg
4 Bremen	12 Köln	20 Bielefeld
5 Essen	13 Mannheim	21 Lübeck
6 Freiburg	14 München	22 Münster
7 Hamburg	15 Nürnberg	
8 Hof	16 Passau	

The results are shown diagrammatically in Fig. B2. The three clusters encircled by continuous lines are the results of NMAX = 3, 4 and RHO = 80, while the six clusters ringed by a broken line are those produced by NMAX = 6,7 and RHO = 70. Fig.B3 contains corresponding results for differently-ordered input sequence of the 22 towns; the differences are evident from studying B3 in conjunction with B2.

Fig. E 1.3

```
M= 22  L= 2  DELTA=  10.0


 -57.0  54.0  46.0   8.0 -36.0 -22.0  34.0  74.0  -6.0  21.0
  37.0 -38.0  -5.0  70.0  59.0 114.0  83.0 -40.0  31.0   0.0
  50.0 -20.0
  28.0 -65.0  79.0 111.0  52.0 -76.0 129.0   6.0 -41.0  45.0
 155.0  35.0 -24.0 -74.0 -26.0 -56.0 -41.0 -28.0 -12.0  71.0
 140.0  70.0
```

		1	2	3	4	5	6	7	8	9	10	11	12	13	14	15	16	17	18	19	20	21	22
3	10.0	1	2	3	0	0	0	0	0	0	0	0	0	0	0	0	0	0	0	0	0	0	0
3	20.0	1	2	3	0	0	0	0	0	0	0	0	0	0	2	0	0	0	0	0	0	0	0
3	30.0	1	2	3	0	0	0	0	0	0	0	0	1	0	2	0	0	0	0	0	0	0	0
3	40.0	1	2	3	0	1	0	0	0	0	0	0	1	0	2	2	0	2	0	0	0	0	0
3	50.0	1	2	3	3	1	0	0	0	0	3	0	1	0	2	2	0	2	0	0	3	0	0
3	60.0	1	2	3	3	1	0	3	0	0	3	0	1	0	2	2	0	2	1	2	3	0	1
3	70.0	1	2	3	3	1	0	3	0	2	3	0	1	0	2	2	2	2	1	2	3	3	1
3	80.0	1	2	3	3	1	2	3	2	2	1	3	1	1	2	2	2	2	1	2	1	3	1
4	10.0	1	2	3	4	0	0	0	0	0	0	0	0	0	0	0	0	0	0	0	0	0	0
4	20.0	1	2	3	4	0	0	0	0	0	0	0	0	0	2	0	0	0	0	0	0	0	0
4	30.0	1	2	3	4	0	0	0	0	0	0	0	1	0	2	0	0	0	0	0	0	0	0
4	40.0	1	2	3	4	1	0	4	0	0	0	0	1	0	2	2	0	2	0	0	0	0	0
4	50.0	1	2	3	3	1	4	0	0	4	3	0	1	0	2	2	0	2	0	0	3	0	0
4	60.0	1	2	3	3	1	4	3	0	4	3	0	1	4	2	2	0	2	1	2	3	0	1
4	70.0	1	2	3	3	1	4	3	0	2	3	0	1	4	2	2	2	2	1	2	3	3	1
4	80.0	1	2	3	3	1	2	3	2	2	1	3	1	1	2	2	2	2	1	2	1	3	1
5	10.0	1	2	3	4	5	0	0	0	0	0	0	0	0	0	0	0	0	0	0	0	0	0
5	20.0	1	2	3	4	5	0	0	0	0	0	0	5	0	2	0	0	0	0	0	0	0	0
5	30.0	1	2	3	4	5	0	0	0	0	0	0	1	0	2	0	0	0	0	0	0	0	5
5	40.0	1	2	3	4	1	5	4	0	5	0	0	1	0	2	2	0	2	0	0	0	0	0
5	50.0	1	2	3	3	1	4	5	0	4	3	5	1	0	2	2	0	2	0	0	3	5	0
5	60.0	1	2	3	3	1	4	3	5	4	3	0	1	4	2	2	0	2	1	2	3	0	1
5	70.0	1	2	3	3	1	4	3	5	2	3	0	1	4	2	2	2	2	1	2	3	3	1
5	80.0	1	2	3	3	1	2	3	2	2	1	3	1	1	2	2	2	2	1	2	1	3	1
6	10.0	1	2	3	4	5	6	0	0	0	0	0	0	0	0	0	0	0	0	0	0	0	0
6	20.0	1	2	3	4	5	6	0	0	0	0	0	5	0	2	0	0	0	0	0	0	0	0
6	30.0	1	2	3	4	5	6	0	0	0	0	0	1	0	2	0	0	0	0	0	0	0	5
6	40.0	1	2	3	4	1	5	4	6	5	0	0	1	0	2	2	0	2	0	0	0	0	0
6	50.0	1	2	3	3	1	4	5	6	4	3	5	1	0	2	2	0	2	0	6	3	5	0
6	60.0	1	2	3	3	1	4	3	5	4	3	6	1	4	2	2	0	2	1	2	3	6	1
6	70.0	1	2	3	3	1	4	3	5	2	3	6	1	4	2	2	2	2	1	2	3	3	1
7	10.0	1	2	3	4	5	6	7	0	0	0	0	0	0	0	0	0	0	0	0	0	0	0
7	20.0	1	2	3	4	5	6	7	0	0	0	0	5	0	2	0	0	0	0	0	0	7	0
7	30.0	1	2	3	4	5	6	7	0	0	0	7	1	0	2	0	0	0	0	0	0	7	5
7	40.0	1	2	3	4	1	5	4	6	5	7	0	1	0	2	2	0	2	0	0	7	0	0
7	50.0	1	2	3	3	1	4	5	6	4	3	5	1	7	2	2	0	2	7	6	3	5	0
7	60.0	1	2	3	3	1	4	3	5	4	3	6	1	4	2	2	7	2	1	2	3	6	1

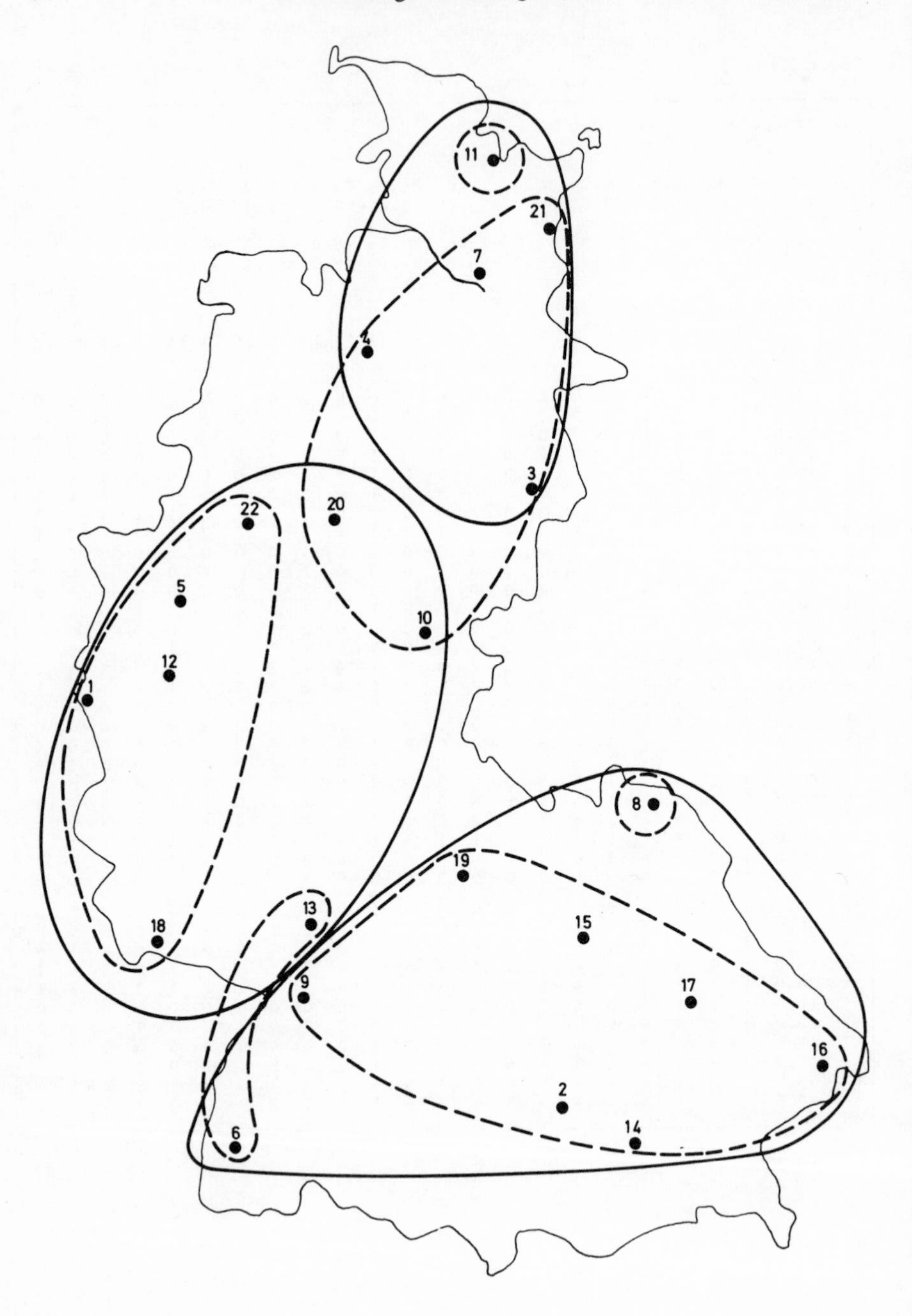

Fig. B 2

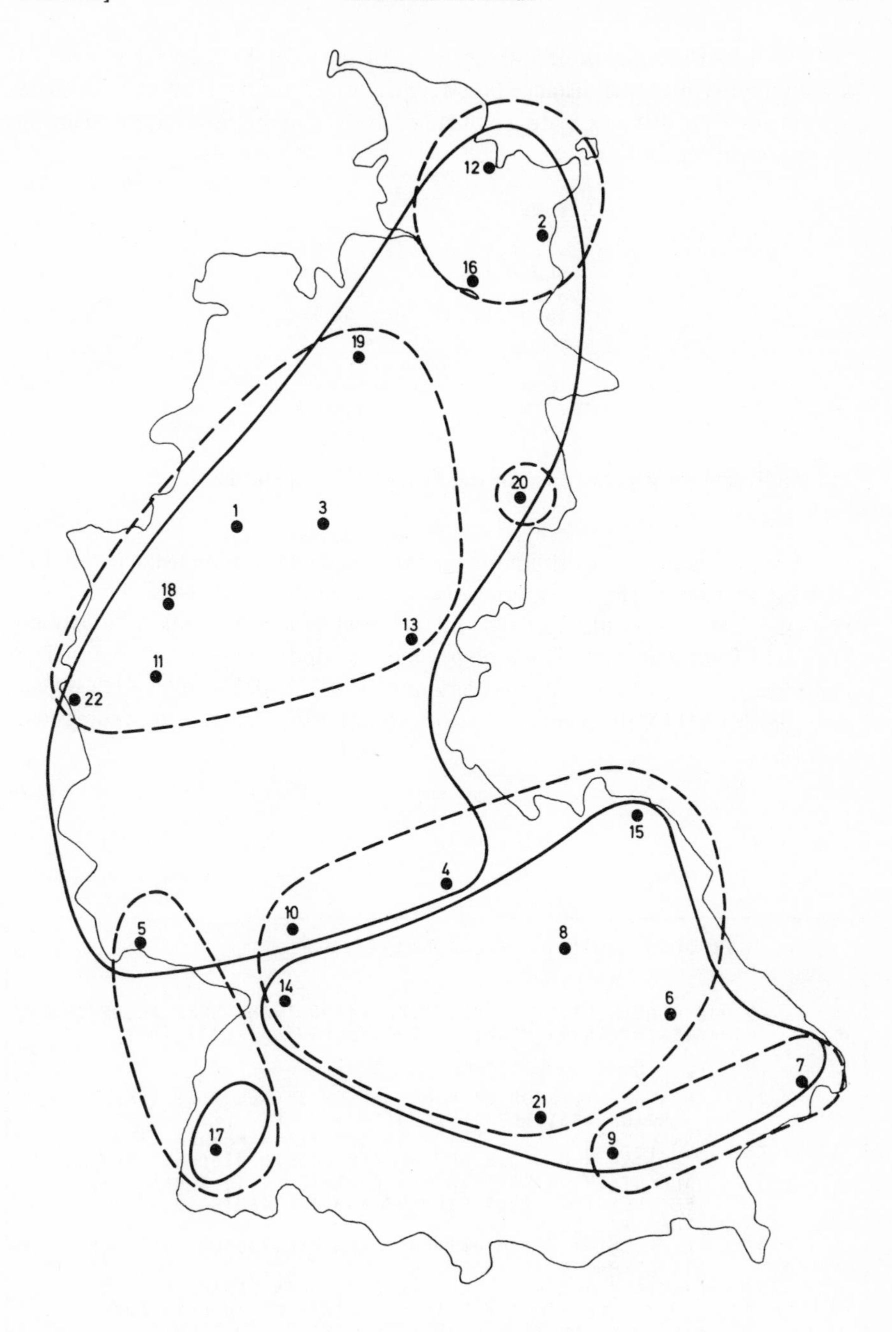

Fig. B 3

The LEADER algorithm, which is given here for Euclidean distances, is easy to modify for other distance functions discussed in Chapter 2. It can also be implemented for distance matrices obtained from non-metric data, by changing the statement sequence

```
      D = 0,
      DO 1 K = 1, L
      T = X(R,K) - X(I,K)
      D = D + T*T
1     CONTINUE
      IF(SQRT(D).GT.RHO) GOTO 2
```

to

```
      IF(D(R,I).GT.RHO) GOTO 2
```

and modifying the argument list of the SUBROUTINE statment.

A second heuristic algorithm for the same data matrix is implemented in the JOINER subroutine (Fig. U3), using the Euclidean metric. As we shall see more explicitly later, it uses the fact that the centroid of $k + 1$ points can easily be calculated from that of the first k points and the additional $(k + 1)$th point. The algorithm is essentially more time-consuming than LEADER, since each object is considered NMAX times and its distance from the centroid has to be computed each time.

Fig. U 3

```
      SUBROUTINE JOINER (M,L,X,NMAX,TAU,P)
C
C     AD HOC CONSTRUCTION OF CLUSTERS WITHOUT ANY SPECIAL OPTIMAL
C     CHARACTERISTICS BY MEANS OF THE FOLLOWING ALGORITHM:
C
C     1.    N = 0
C     2.    N = N + 1, STOP IF N > NMAX OR IF ALL X(I) ARE
C           ALREADY ASSIGNED TO CLUSTERS.
C     3.    AN ELEMENT X(I), WHICH IS NOT YET A MEMBER OF A
C           CLUSTER AND WHOSE SQUARED EUCLIDEAN DISTANCE FROM
C           ALL OTHER OBJECTS AS YET UNASSIGNED IS A MAXIMUM,
C           BECOMES THE FIRST ELEMENT AND THE CENTROID OF
C           THE N-TH CLUSTER.
C     4.    AN ELEMENT WHICH HAS NOT BEEN PREVIOUSLY ASSIGNED
C           IS LIKEWISE ASSIGNED TO THE N-TH CLUSTER IF IT IS
C           THE ONE WHICH IS, IN THE EUCLIDEAN SENSE, CLOSEST
C           TO THE CENTROID AND IF THE DISTANCE IS LESS THAN
C           OR EQUAL TO TAU; OTHERWISE, RETURN TO STEP 2.
C     5.    THE CENTROID IS RECALCULATED AND THE PROCESS IN
C           STEP 4 IS REPEATED.
```

```
C
C     P(I) CONTAINS THE NUMBER OF THE CLUSTER TO WHICH X(I)
C     HAS BEEN ASSIGNED,   P(I) = 0 MEANS NO ASSIGNMENT,
C
C     DIMENSION X(M,L),   P(M),   S(NMAX,L),Q(NMAX)
      DIMENSION X(100,12),P(100),S(10,12) ,Q(10)
      INTEGER P,R,U,Q
      DO 1 I=1,M
           P(I)=0
    1 CONTINUE
      M1=M-1
      N=0
    2 N=N+1
      IF(N.GT.NMAX) RETURN
      R=0
      U=0
      H=0.
      DO 5 I=1,M1
           IF(P(I).NE.0) GOTO 5
           U=I
           I1=I+1
           DO 4 J=I1,M
                IF(P(J).NE.0) GOTO 4
                F=0.
                DO 3 K=1,L
                     T=X(I,K)-X(J,K)
                     F=F+T*T
    3           CONTINUE
                IF(F.LE.H) GOTO 4
                H=F
                R=I
    4      CONTINUE
    5 CONTINUE
      IF(R.NE.0) GOTO 6
      IF(U.NE.0) P(U)=N
      RETURN
    6 P(R)=N
      Q(N)=1
      DO 7 K=1,L
           S(N,K)=X(R,K)
    7 CONTINUE
    8 H=1.E30
      DO 10 I=1,M
            IF(P(I).NE.0) GOTO 10
            F=0.
            DO 9 K=1,L
                 T=S(N,K)-X(I,K)
                 F=F+T*T
    9       CONTINUE
            IF(F.GE.H) GOTO 10
            H=F
            R=I
   10 CONTINUE
      IF(SQRT(H).GT.TAU) GOTO 2
      P(R)=N
      U=Q(N)
      V=FLOAT(U)
      W=1./(V+1.)
      Q(N)=U+1
      DO 11 K=1,L
            S(N,K)=W*(V*S(N,K)+X(R,K))
```

```
   11 CONTINUE
      GOTO 8
      END
```

Fig. U 3

The main program H2 is almost identical to H1; the tablets of results in Fig. E2.1 to E2.3 correspond to Fig. E1.1 to E1.3.

```
C
C     TEST PROGRAM FOR THE JOINER ALGORITHM
C
      DIMENSION X(100,12),P(100)
      INTEGER P
      KI=5
      KO=6
    1 READ(KI,2) M,L,N1,N2,DELTA
    2 FORMAT(4I5,F5.0)
      IF(M.LE.0.OR.M.GT.100.OR.L.LT.1.OR.L.GT.12
     *   .OR.N1.LE.1.OR.N2.GT.10) STOP
      IF(DELTA.LE.0.) DELTA=1.
      WRITE(KO,3) M,L,DELTA
    3 FORMAT('1',' M=',I3,' L=',I2,'  DELTA=',F6.1)
      WRITE(KO,4)
    4 FORMAT('0')
      DO 5 K=1,L
            READ(KI,6)  (X(I,K),I=1,M)
            WRITE(KO,7) (X(I,K),I=1,M)
    5 CONTINUE
      WRITE(KO,4)
      WRITE(KO,11) (I,I=1,M)
      WRITE(KO,4)
    6 FORMAT(16F5.0)
    7 FORMAT(3X,10F6.1)
      N=N1
      RHO=0.
    8 RHO=RHO+DELTA
      CALL JOINER (M,L,X,N,RHO,P)
      WRITE(KO,9) N,RHO,(P(I),I=1,M)
    9 FORMAT(1X,I3,F6.1,23I3/(15X,23I3))
      DO 10 I=1,M
            IF(P(I).EQ.0) GOTO 8
   10 CONTINUE
   11 FORMAT(10X,23I3)
      RHO=0.
      N=N+1
      IF(N.LE.N2) GOTO 8
      GOTO 1
      END
```

Fig. H 2

M= 15 L= 1 DELTA= 0.3

0.0 0.1 0.2 0.5 2.0 2.1 2.5 3.0 3.2 4.0
5.0 6.0 8.0 10.0 11.0

		1	2	3	4	5	6	7	8	9	10	11	12	13	14	15
3	0.3	1	1	1	2	3	3	0	0	0	0	0	0	0	0	0
3	0.6	1	1	1	1	2	2	2	3	3	0	0	0	0	0	0
3	0.9	1	1	1	1	2	2	2	2	2	3	0	0	0	0	0
3	1.2	1	1	1	1	2	2	2	2	2	3	3	0	0	0	0
3	1.5	1	1	1	1	2	2	2	2	2	2	3	3	0	0	0
3	1.8	1	1	1	1	2	2	2	2	2	2	3	3	0	0	0
3	2.1	1	1	1	1	1	1	1	1	1	2	2	2	3	3	3
4	0.3	1	1	1	2	3	3	4	0	0	0	0	0	0	0	0
4	0.6	1	1	1	1	2	2	2	3	3	4	0	0	0	0	0
4	0.9	1	1	1	1	2	2	2	2	2	3	4	0	0	0	0
4	1.2	1	1	1	1	2	2	2	2	2	3	3	4	0	0	0
4	1.5	1	1	1	1	2	2	2	2	2	2	3	3	4	0	0
4	1.8	1	1	1	1	2	2	2	2	2	2	3	3	4	0	0
4	2.1	1	1	1	1	1	1	1	1	1	2	2	2	3	3	3
5	0.3	1	1	1	2	3	3	4	5	5	0	0	0	0	0	0
5	0.6	1	1	1	1	2	2	2	3	3	4	5	0	0	0	0
5	0.9	1	1	1	1	2	2	2	2	2	3	4	5	0	0	0
5	1.2	1	1	1	1	2	2	2	2	2	3	3	4	5	0	0
5	1.5	1	1	1	1	2	2	2	2	2	2	3	3	4	5	5
6	0.3	1	1	1	2	3	3	4	5	5	6	0	0	0	0	0
6	0.6	1	1	1	1	2	2	2	3	3	4	5	6	0	0	0
6	0.9	1	1	1	1	2	2	2	2	2	3	4	5	6	0	0
6	1.2	1	1	1	1	2	2	2	2	2	3	3	4	5	6	6
7	0.3	1	1	1	2	3	3	4	5	5	6	7	0	0	0	0
7	0.6	1	1	1	1	2	2	2	3	3	4	5	6	7	0	0
7	0.9	1	1	1	1	2	2	2	2	2	3	4	5	6	7	0
7	1.2	1	1	1	1	2	2	2	2	2	3	3	4	5	6	6
8	0.3	1	1	1	2	3	3	4	5	5	6	7	8	0	0	0
8	0.6	1	1	1	1	2	2	2	3	3	4	5	6	7	8	0
8	0.9	1	1	1	1	2	2	2	2	2	3	4	5	6	7	0
8	1.2	1	1	1	1	2	2	2	2	2	3	3	4	5	6	6
9	0.3	1	1	1	2	3	3	4	5	5	6	7	8	9	0	0
9	0.6	1	1	1	1	2	2	2	3	3	4	5	6	7	8	0
9	0.9	1	1	1	1	2	2	2	2	2	3	4	5	6	7	0
9	1.2	1	1	1	1	2	2	2	2	2	3	3	4	5	6	6
10	0.3	1	1	1	2	3	3	4	5	5	6	7	8	9	10	0
10	0.6	1	1	1	1	2	2	2	3	3	4	5	6	7	8	0
10	0.9	1	1	1	1	2	2	2	2	2	3	4	5	6	7	0
10	1.2	1	1	1	1	2	2	2	2	2	3	3	4	5	6	6

Fig. E 2.1

```
M= 15   L= 1   DELTA=    0.3

  11.0    2.5    8.0    3.2    5.0    0.1   10.0    2.1    3.0    0.5
   2.0    6.0    0.0    0.2    4.0
```

		1	2	3	4	5	6	7	8	9	10	11	12	13	14	15
3	0.3	1	0	3	0	0	0	2	0	0	0	0	0	0	0	0
3	0.6	1	0	3	0	0	0	2	0	0	0	0	0	0	0	0
3	0.9	1	0	3	0	0	0	2	0	0	0	0	0	0	0	0
3	1.2	1	0	2	0	3	0	1	0	0	0	0	3	0	0	0
3	1.5	1	0	2	0	3	0	1	0	0	0	0	3	0	0	0
3	1.8	1	0	2	0	3	0	1	0	0	0	0	3	0	0	3
3	2.1	1	3	2	3	2	3	1	3	3	3	3	2	3	3	0
3	2.4	1	3	2	3	2	3	1	3	3	3	3	2	3	3	2
4	0.3	1	0	3	0	0	0	2	0	0	0	0	4	0	0	0
4	0.6	1	0	3	0	0	0	2	0	0	0	0	4	0	0	0
4	0.9	1	0	3	0	0	0	2	0	0	0	0	4	0	0	0
4	1.2	1	0	2	0	3	4	1	0	0	4	0	3	4	4	0
4	1.5	1	0	2	0	3	4	1	0	0	4	0	3	4	4	0
4	1.8	1	4	2	4	3	0	1	4	4	0	4	3	0	0	3
4	2.1	1	3	2	3	2	3	1	3	3	3	3	2	3	3	0
4	2.4	1	3	2	3	2	3	1	3	3	3	3	2	3	3	2
5	0.3	1	0	3	0	5	0	2	0	0	0	0	4	0	0	0
5	0.6	1	0	3	0	5	0	2	0	0	0	0	4	0	0	0
5	0.9	1	0	3	0	5	0	2	0	0	0	0	4	0	0	0
5	1.2	1	5	2	5	3	4	1	5	5	4	5	3	4	4	0
5	1.5	1	5	2	5	3	4	1	5	5	4	5	3	4	4	5
6	0.3	1	0	3	0	5	6	2	0	0	0	0	4	6	6	0
6	0.6	1	0	3	0	5	6	2	0	0	6	0	4	6	6	0
6	0.9	1	0	3	0	5	6	2	0	0	6	0	4	6	6	0
6	1.2	1	5	2	5	3	4	1	5	5	4	5	3	4	4	0
6	1.5	1	5	2	5	3	4	1	5	5	4	5	3	4	4	5
7	0.3	1	0	3	0	5	6	2	0	0	7	0	4	6	6	0
7	0.6	1	7	3	0	5	6	2	7	0	6	7	4	6	6	0
7	0.9	1	7	3	7	5	6	2	7	7	6	7	4	6	6	0
7	1.2	1	5	2	5	3	4	1	5	5	4	5	3	4	4	0
7	1.5	1	5	2	5	3	4	1	5	5	4	5	3	4	4	5
8	0.3	1	0	3	0	5	6	2	8	0	7	8	4	6	6	0
8	0.6	1	7	3	8	5	6	2	7	8	6	7	4	6	6	0
8	0.9	1	7	3	7	5	6	2	7	7	6	7	4	6	6	0
8	1.2	1	5	2	5	3	4	1	5	5	4	5	3	4	4	0
8	1.5	1	5	2	5	3	4	1	5	5	4	5	3	4	4	5
9	0.3	1	9	3	0	5	6	2	8	0	7	8	4	6	6	0
9	0.6	1	7	3	8	5	6	2	7	8	6	7	4	6	6	0
9	0.9	1	7	3	7	5	6	2	7	7	6	7	4	6	6	0
9	1.2	1	5	2	5	3	4	1	5	5	4	5	3	4	4	0
9	1.5	1	5	2	5	3	4	1	5	5	4	5	3	4	4	5
10	0.3	1	9	3	10	5	6	2	8	10	7	8	4	6	6	0
10	0.6	1	7	3	8	5	6	2	7	8	6	7	4	6	6	0
10	0.9	1	7	3	7	5	6	2	7	7	6	7	4	6	6	0
10	1.2	1	5	2	5	3	4	1	5	5	4	5	3	4	4	0
10	1.5	1	5	2	5	3	4	1	5	5	4	5	3	4	4	5

Fig. E 2.2

```
M= 22   L= 2   DELTA=   10.0

 -57.0  54.0  46.0   8.0 -36.0 -22.0  34.0  74.0  -6.0  21.0
  37.0 -38.0  -5.0  70.0  59.0 114.0  83.0 -40.0  31.0   0.0
  50.0 -20.0
  28.0 -65.0  79.0 111.0  52.0 -76.0 129.0   6.0 -41.0  45.0
 155.0  35.0 -24.0 -74.0 -26.0 -56.0 -41.0 -28.0 -12.0  71.0
 140.0  70.0
```

		1	2	3	4	5	6	7	8	9	10	11	12	13	14	15	16	17	18	19	20	21	22
3	10.0	0	0	0	0	0	1	0	0	0	0	2	0	0	3	0	0	0	0	0	0	0	0
3	20.0	0	0	0	0	0	1	3	0	0	0	2	0	0	0	0	0	0	0	0	0	2	0
3	30.0	0	0	0	3	0	1	2	0	0	0	2	0	0	0	0	0	0	0	0	0	2	0
3	40.0	0	0	0	3	0	1	2	0	1	0	2	0	1	0	0	0	0	1	0	0	2	0
3	50.0	3	0	0	2	3	1	2	0	1	0	2	3	1	0	0	0	0	1	0	3	2	3
3	60.0	3	0	2	2	3	1	2	0	1	3	2	3	1	0	0	0	0	1	1	3	2	3
3	70.0	3	1	2	2	3	1	2	1	1	2	2	3	1	1	1	0	1	1	1	2	2	2
3	80.0	3	1	2	2	2	1	2	1	1	2	2	3	1	1	1	0	1	1	1	2	2	2
3	90.0	2	1	2	2	2	1	2	1	1	1	2	2	1	1	1	1	1	1	1	2	2	2
4	10.0	0	0	0	0	0	1	0	0	0	0	2	0	0	3	0	4	0	0	0	0	0	0
4	20.0	0	0	0	4	0	1	3	0	0	0	2	0	0	0	0	0	0	0	0	0	2	0
4	30.0	4	0	0	3	4	1	2	0	0	0	2	4	0	0	0	0	0	0	0	0	2	0
4	40.0	4	0	0	3	4	1	2	0	1	0	2	4	1	0	0	0	0	1	0	0	2	4
4	50.0	3	0	4	2	3	1	2	0	1	4	2	3	1	0	0	0	0	1	0	3	2	3
4	60.0	3	4	2	2	3	1	2	4	1	3	2	3	1	4	4	4	4	1	1	3	2	3
5	10.0	0	5	0	0	0	1	0	0	0	0	2	0	0	3	0	4	0	0	0	0	0	0
5	20.0	5	0	0	4	0	1	3	0	0	0	2	0	0	0	0	0	0	0	0	0	2	0
5	30.0	4	0	0	3	4	1	2	0	0	0	2	4	0	0	0	5	0	0	0	0	2	0
5	40.0	4	5	0	3	4	1	2	0	1	0	2	4	1	5	5	5	5	1	0	0	2	4
5	50.0	3	5	4	2	3	1	2	0	1	4	2	3	1	5	5	5	5	1	0	3	2	3
5	60.0	3	4	2	2	3	1	2	4	1	3	2	3	1	4	4	4	4	1	1	3	2	3
6	10.0	0	5	0	0	0	1	0	0	0	0	2	0	0	3	0	4	0	6	0	0	0	0
6	20.0	5	0	0	4	6	1	3	0	0	0	2	6	0	0	0	0	0	0	0	0	2	0
6	30.0	4	6	0	3	4	1	2	0	0	0	2	4	0	6	0	5	0	0	0	0	2	0
6	40.0	4	5	0	3	4	1	2	6	1	0	2	4	1	5	5	5	5	1	0	0	2	4
6	50.0	3	5	4	2	3	1	2	6	1	4	2	3	1	5	5	5	5	1	6	3	2	3
7	10.0	0	5	0	0	0	1	0	0	7	0	2	0	0	3	0	4	0	6	0	0	0	0
7	20.0	5	0	0	4	6	1	3	0	0	0	2	6	0	0	0	7	0	0	0	0	2	0
7	30.0	4	6	0	3	4	1	2	0	0	0	2	4	0	6	7	5	7	0	0	0	2	0
7	40.0	4	5	7	3	4	1	2	6	1	0	2	4	1	5	5	5	5	1	0	0	2	4
7	50.0	3	5	4	2	3	1	2	6	1	4	2	3	1	5	5	5	5	1	6	3	2	3

Fig. E 2.3

Before considering further algorithms, we shall show how JOINER can be used without calculating the centroid and thus can use, instead of a metric data matrix, a distance matrix which may not have been produced by the use of the Euclidean metric.

In what follows we shall frequently denote the ith row of a data matrix by $x_i = x_{i.}^T$ and the centroid (mean) vector by

$$\bar{x} = \frac{1}{m} \sum_{i=1}^{m} x_i \tag{3.1.4}$$

On the basis of this definition,

$$\sum_{i=1}^{m} \|x_i - \bar{x}\|^2 \quad (\|.\| = \|.\|_2) \tag{3.1.5}$$

is minimised and

$$\sum_{i=1}^{m} (x_i - \bar{x}) = 0. \tag{3.1.6}$$

is valid. Let y be an arbitrary real vector with l components. The expression

$$\sum_{i=1}^{m} \|x_i - y\|^2 = \sum_{i=1}^{m} \|x_i - \bar{x}\|^2 + m\,\|\bar{x} - y\|^2. \tag{3.1.7}$$

then holds. The proof runs as follows:

$$x_i - y = (x_i - \bar{x}) + (\bar{x} - y)$$

$$\|x_i - y\|^2 = \|x_i - \bar{x}\|^2 + 2\,(x_i - \bar{x})^T (\bar{x} - y) + \|\bar{x} - y\|^2$$

$$\sum_{i=1}^{m} \|x_i - y\|^2 = \sum_{i=1}^{m} \|x_i - \bar{x}\|^2 + m\,\|\bar{x} - y\|^2$$

$$+ 2 \sum_{i=1}^{m} (\bar{x} - y)^T (x_i - \bar{x}).$$

The last term on the right-hand side vanishes as a consequence of (3.1.6). Since the terms on the right-hand side of (3.1.7) are non-negative, it is clear that the left-hand side is a minimum for $y = \bar{x}$.

By setting $y = x_k$ in (3.1.7) and summing over $k = 1, \ldots, m$ a relationship is obtained which is very important for the whole of Chapter 3:

$$\sum_{i=1}^{m} \|x_i - \bar{x}\|^2 = \frac{1}{2m} \sum_{i=1}^{m} \sum_{k=1}^{m} \|x_i - x_k\|^2 \tag{3.1.8}$$

$$= \frac{1}{m} \sum_{i=1}^{m} \sum_{k>i} \|x_i - x_k\|^2.$$

Rewriting (3.1.8) for a cluster characterised by C_j, we obtain

$$\sum_{i \in C_j} \|x_i - \bar{x}_j\|^2 = \frac{1}{2m_j} \sum_{i \in C_j} \sum_{k \in C_j} \|x_i - x_k\|^2, \tag{3.1.9}$$

in which $\bar{x}_j$ is the mean vector of the cluster C_j. This result means that the sum of the squared Euclidean distances of the members of the cluster from their centroid can be expressed in terms of the distances between the individual members. By considering the left-hand side when written as

$$\frac{1}{2m_j} \sum_{i \in C_j} \sum_{k \in C_j} d^2(x_i, x_j), \tag{3.1.10}$$

where d is the Euclidean metric, it becomes clear how the JOINER algorithm can work when using a precalculated matrix $d(x_i, x_j)$. We shall show this explicitly in the algorithm CLUDIA which comes later.

It is even possible to use a matrix of non-metric distances in (3.1.10) rather than a Euclidean distance matrix. This use is naturally not based on (3.1.9), which holds only for the Euclidean metric, but it is legitimate if the corresponding cluster algorithm is successful. (See Bock 1974, Elton and Gruber 1970, Frank and Green 1968, Green, Frank and Robinson 1967, Kendall 1966 and Sneath 1966).

In addition to the use of a threshold value, the JOINER algorithm can also (see Green and Carmone 1970) be implemented with a fixed maximum number of objects which may be assigned to a cluster. In addition, a new cluster can always be started by using those two objects which are closest together, rather than sometimes selecting the two that are furthest away.

Because of the speed of modern computers, heuristic algorithms are mainly used as pre-processors for partitioning procedures involving objective functions, even for large values of m. These partitioning techniques are also useful without the preceding use of heuristic algorithms, as the remainder of this chapter will show.

We wish, nevertheless, to discuss still further the formation of precisely two clusters with a certain algorithm (see Edwards and Cavalli-Sforza 1965, Bock 1970). This is given in the ZWEIGO subroutine in Fig. U4 and is applicable to metric matrices only. The arrays S1(L) and S2(L) contain the coordinates of the centroids found; DS contains the Euclidean distance between these centroids.

Fig. U 4

```
      SUBROUTINE ZWEIGO (M,L,X,P,S1,S2,DS,IDR)
C
C     ALGORITHM:
C
C     1.    INITIALLY THE TWO VECTORS X WHOSE DISTANCE APART IS
C           A MAXIMUM ARE TAKEN AS THE CENTROIDS S1 AND S2.
C           TWO CLUSTERS OF ONE ELEMENT ARE OBTAINED.
C     2.    FROM THE VECTORS X AS YET UNASSIGNED (P(I) = 0),
C           THOSE WITH THE SMALLEST DISTANCES D1 FROM S1 AND
C           D2 FROM S2 ARE CHOSEN, AS X1 AND X2 RESPECTIVELY.
C     3.    IF D1 < D2, X1 IS ASSIGNED TO THE FIRST CLUSTER;
C           OTHERWISE X2 IS ASSIGNED TO THE SECOND CLUSTER.
C           THE CORRESPONDING CENTROID IS MODIFIED.
```

```
C
C     STEPS 2 AND 3 ARE REPEATED UNTIL ALL M VECTORS ARE
C     ASSIGNED,
C
C     IF IDR IS NON-ZERO, THE CURRENT VALUES OF THE VECTOR P
C     AND OF DS ARE PRINTED AT EACH ITERATION,
C
C     DIMENSION X(M,L),    P(M),   S1(L), S2(L)
      DIMENSION X(100,12),P(100),S1(12),S2(12)
      INTEGER P,G1,G2
      KO=6
      IF(M.LE.2) RETURN
      M1=M-1
      DO 1 I=1,M
           P(I)=0
    1 CONTINUE
      DMAX=0.
      DO 4 I=1,M1
           I1=I+1
           DO 3 J=I1,M
                H=0.
                DO 2 K=1,L
                     F=X(I,L)-X(J,L)
                     H=H+F*F
    2           CONTINUE
                IF(H.LE.DMAX) GOTO 3
                DMAX=H
                IC=I
                JC=J
    3      CONTINUE
    4 CONTINUE
      DO 5 K=1,L
           S1(K)=X(IC,K)
           S2(K)=X(JC,K)
    5 CONTINUE
      P(IC)=1
      P(JC)=2
      G1=1
      G2=1
      IP=2
    6 H=0.
      DC 7 K=1,L
           F=S1(K)-S2(K)
           H=H+F*F
    7 CONTINUE
      DS=SQRT(H)
      IF(IDR.EQ.0) GOTU 10
      WRITE(KO,8) (P(I),I=1,M)
    8 FORMAT('0',6X,15I3/(7X,15I3))
      WRITE(KC,9) DS
    9 FORMAT('+',50X,F15.2)
   10 IP=IP+1
      IF(IP.GT.M) RETURN
      D1=1.E30
      D2=D1
      DO 13 I=1,M
            IF(P(I).NE.0) GOTO 13
            H1=0.
            H2=0.
            DC 11 K=1,L
```

```
                H=X(I,K)
                F=S1(K)-H
                H1=H1+F*F
                F=S2(K)-H
                H2=H2+F*F
11          CCNTINUE
            IF(H1.GE.D1) GCTC 12
            D1=H1
            I1=I
12          IF(H2.GE.D2) GCTC 13
            D2=H2
            I2=I
13    CONTINUE
      IF(D1.GE.D2) GOTO 15
      P(I1)=1
      H=FLOAT(G1)
      F=1./(H+1.)
      DO 14 K=1,L
            S1(K)=(H*S1(K)+X(I1,K))*F
14    CCNTINUE
      G1=G1+1
      GOTO 6
15    P(I2)=2
      H=FLOAT(G2)
      F=1./(H+1.)
      DC 16 K=1,L
            S2(K)=(H*S2(K)+X(I2,K))*F
16    CONTINUE
      G2=G2+1
      GCTO 6
      END
```

Fig. H 3

```
C
C      HEURISTIC PARTITIONING INTO TWO CLUSTERS
C
       DIMENSION X(100,12),P(100),S1(12),S2(12)
       INTEGER P
       KI=5
       KO=6
     1 READ(KI,2) M,L,N1,N1,IDR
     2 FORMAT(16I5)
       IF(M.LE.2.OR.M.GT.10C.OR.L.LT.1.OR.L.GT.12) STOP
       WRITE(KO,3) M,L,IDR
     3 FORMAT('1',' M=',I3,'  L=',I2,'  IDR=',I1)
       WRITE(KO,4)
     4 FORMAT('0')
       DO 5 K=1,L
             READ(KI,6)  (X(I,K),I=1,M)
             WRITE(KO,7) (X(I,K),I=1,M)
     5 CONTINUE
     6 FORMAT(16F5.0)
```

```
    7 FORMAT(1X,10F7.1)
      WRITE(KO,4)
      CALL ZWEIGO (M,L,X,P,S1,S2,DS,IDR)
      WRITE(KO,4)
      WRITE(KO,8) (P(I),I=1,M)
    8 FORMAT(7X,15I3)
      WRITE(KO,4)
      WRITE(KO,9) (S1(K),K=1,L)
      WRITE(KO,9) (S2(K),K=1,L)
    9 FORMAT(1X,5F13.2)
      WRITE(KO,4)
      WRITE(KO,9) DS
      GOTO 1
      END
```

Fig. H 3

Fig. E3.1 shows the application of this algorithm, by means of the program shown in Fig. H3, to the coordinates of the towns in Fig. T1. A further example, which will similarly be used frequently, consists of the M = 59 towns shown in Fig. B4, whose Cartesian co-ordinates, as will be seen more explicitly later, are used as data matrices. The allocation found by ZWEIGO and the corresponding centroids are shown in Fig. B5.

```
M= 22   L= 2   IDR=0

-57.0   54.0   46.0    8.0  -36.0  -22.0   34.0   74.0   -6.0   21.0
 37.0  -38.0   -5.0   70.0   59.0  114.0   83.0  -40.0   31.0    0.0
 50.0  -20.0
 28.0  -65.0   79.0  111.0   52.0  -76.0  129.0    6.0  -41.0   45.0
155.0   35.0  -24.0  -74.0  -26.0  -56.0  -41.0  -28.0  -12.0   71.0
140.0   70.0


        2  1  2  2  2  1  2  1  1  2  2  2  1  1  1
        1  1  1  1  2  2  2


        37.45          -39.73
         4.09           83.18


       127.36
```

Fig. E 3.1

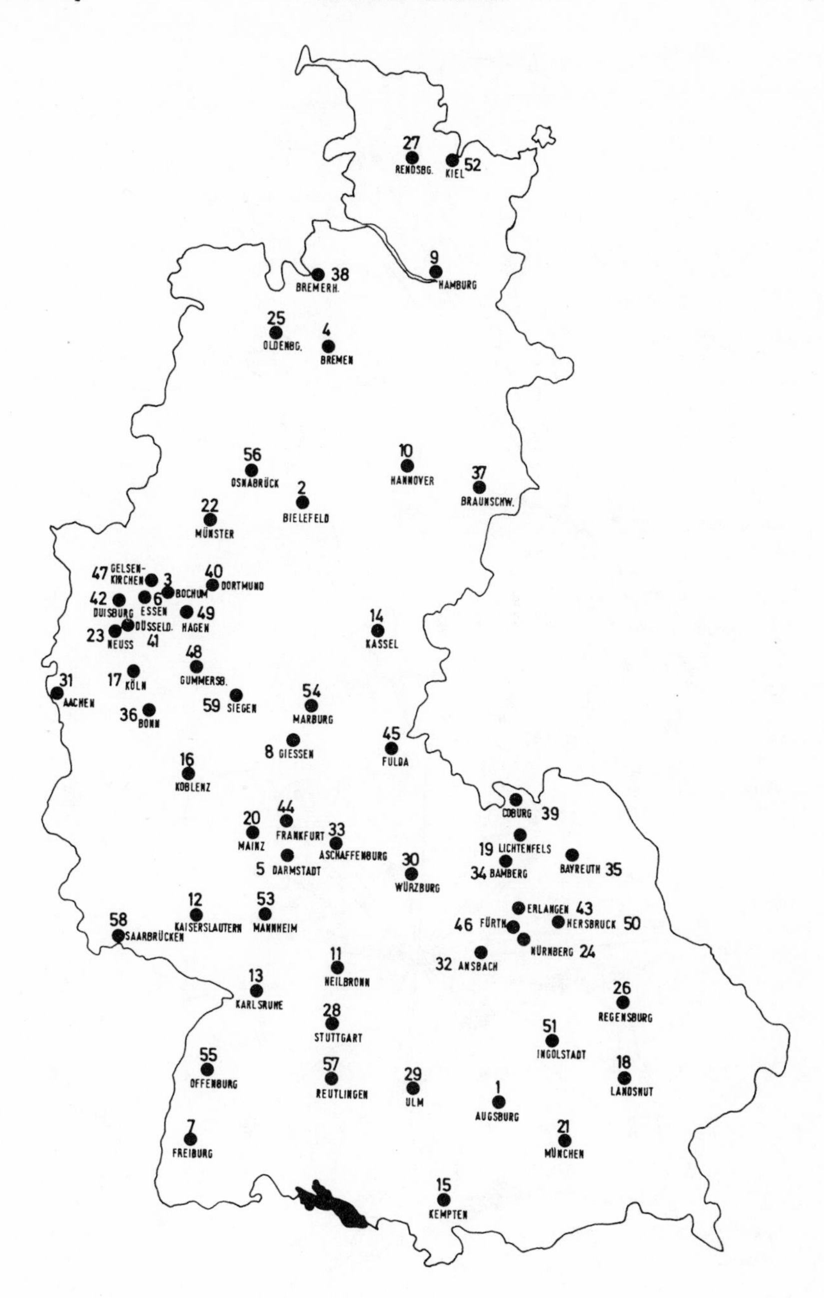
27
RENDSBG.
KIEL
52
9
HAMBURG
38
BREMERH.
25
OLDENBG.
4
BREMEN
56
OSNABRÜCK
10
HANNOVER
37
BRAUNSCHW.
2
BIELEFELD
22
MÜNSTER
47 GELSEN-KIRCHEN
3
40
DORTMUND
42
6
BOCHUM
DUISBURG
ESSEN
49
DÜSSELD.
HAGEN
23
NEUSS
41
14
KASSEL
48
GUMMERSB.
31
AACHEN
17
KÖLN
59
SIEGEN
54
MARBURG
36
BONN
45
FULDA
8
GIESSEN
16
KOBLENZ
COBURG
39
20
MAINZ
44
FRANKFURT
33
ASCHAFFENBURG
19
LICHTENFELS
5
DARMSTADT
30
WÜRZBURG
34
BAMBERG
BAYREUTH
35
12
KAISERSLAUTERN
53
MANNHEIM
ERLANGEN
43
58
SAARBRÜCKEN
46
FÜRTH
HERSBRUCK
50
NÜRNBERG
24
11
HEILBRONN
32
ANSBACH
13
KARLSRUHE
26
REGENSBURG
28
STUTTGART
51
INGOLSTADT
55
OFFENBURG
57
REUTLINGEN
29
ULM
18
LANDSHUT
1
AUGSBURG
7
FREIBURG
21
MÜNCHEN
15
KEMPTEN

Fig. B 4

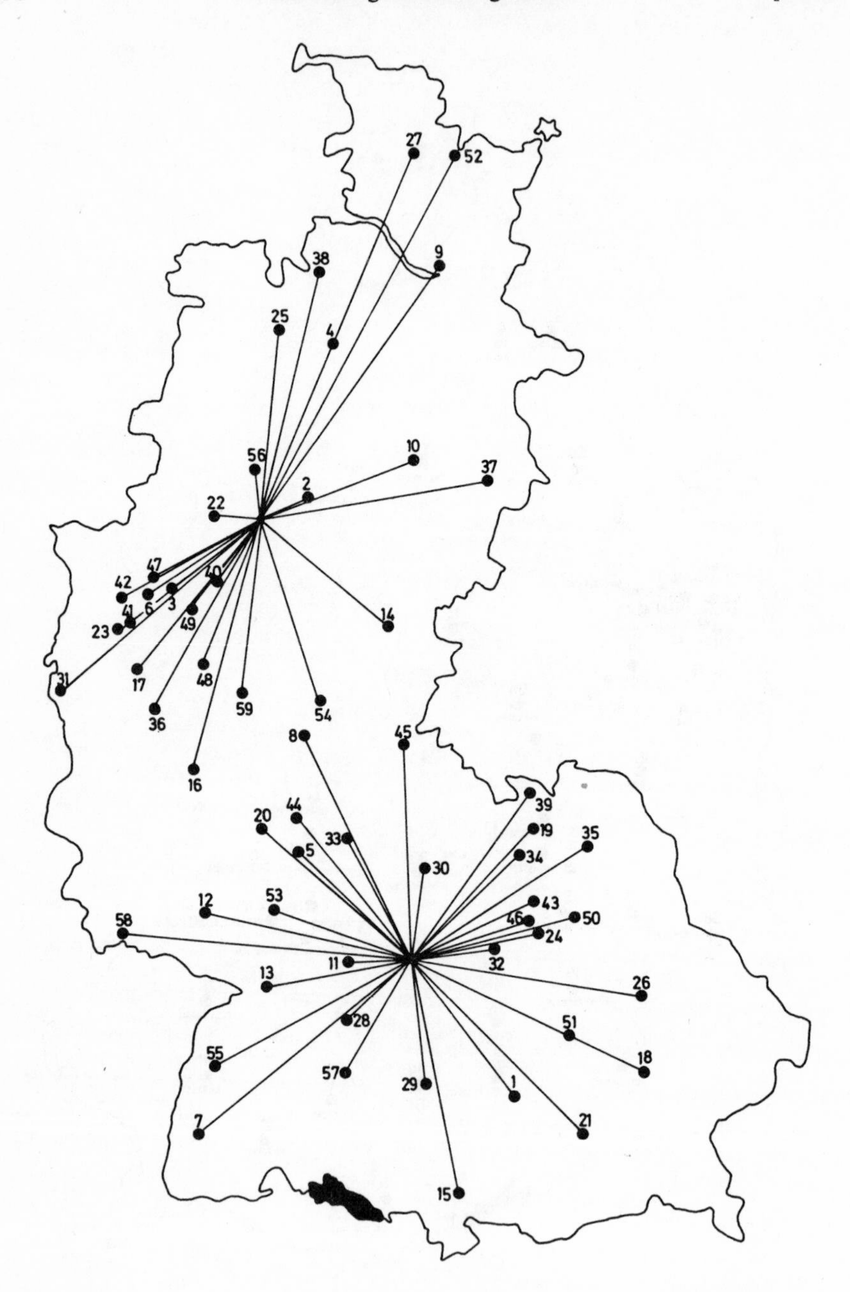
27
52
9
38
25
4
56
10
37
2
22
47
42
40
3
6
41
23
49
14
31
17
48
59
54
36
16
8
45
39
44
20
19
35
33
5
34
30
12
53
43
58
46
50
24
11
32
13
26
28
51
55
18
57
29
1
7
21
15

Fig. B 5

We show by a further example (see Elton and Gruber 1970) that ZWEIGO does not necessarily maximise DS, the distance between the two centroids. Let L = 1, M = 20 and

$$\begin{aligned} x_i &= i-1 \quad (i=1,\dots,7) \\ x_i &= i \quad\quad (i=8,\dots,20) \end{aligned} \tag{3.1.11}$$

As can be established (see Fisher 1958) by enumerating all of the 19 possible partitionings C_1 and C_2 which are significant (i.e. those which are contiguous), the centroid distance DS is maximised for $C_1 = \{1,\dots,7\}$ and $C_2 = \{8,\dots,20\}$, which is what would be expected intuitively. This corresponds to DS = 11, $\bar{x}_1 = 3$ and $\bar{x}_2 = 14$. However, the ZWEIGO subroutine produces the assignment $C_1 = \{1,\dots,9\}$ and $C_2 = \{10,\dots,20\}$ with $\bar{x}_1 = 4.222$, $\bar{x}_2 = 15$ and DS = 10.778.

3.2 THE SUM OF SQUARED DISTANCE CRITERION.

The number $S(m,n)$ of possible partitionings of length n of the m objects represented by the symbols $1,\dots,n$, in accordance with (3.1.1) and (3.1.2), is given by

$$S(m,n) = \frac{1}{n!}\sum_{j=1}^{n}(-1)^{n-j}\binom{n}{j}j^m. \tag{3.2.1}$$

(see Anderberg 1973, Jensen 1969, Lüneburg 1971). $S(m,n)$ obeys the recursion recursion

$$\begin{aligned} S(m,n) &= n\,S(m-1,n) + S(m-1,n-1) \\ \text{with } S(1,1) &= 1 \quad \text{and} \quad S(1,n) = 0 \text{ for } n \neq 1. \end{aligned} \tag{3.2.2}$$

(see Rubin 1967). Accordingly, for the case of two clusters considered in the last section,

$$S(m,2) = 2^{m-1} - 1 \tag{3.2.3}$$

and for the example of 59 towns in E3.1,

$$S(59,2) \approx 10^{18}.$$

Further examples of numerical values are

$$\begin{aligned} S(15,3) &= 2,375,101 \\ S(20,4) &= 45,232,115,901 \\ S(25,8) &= 690,223,721,118,368,580 \\ S(100,5) &= 10^{68}. \end{aligned}$$

(see Bock 1974, Duran and Odell 1974).

It is clear from these examples that the partitioning which is optimal with respect to any objective function cannot be found within a reasonable computation time by combinatorial enumeration, especially since values $m \approx 2000$ and $n \approx 10$ are not unusual in practice.

We now introduce a special objective function, which we are going to examine thoroughly in this section, and for which we take the data matrix (1.1) as given. We recollect our notation $x_i = x_{i.}^T$, by which we mean the ith row of the data matrix, by which the ith object is described in terms of the l variables.

The m vectors x_i contain the information at its most detailed; the centroid

$$\bar{x} = \frac{1}{m} \sum_{i=1}^{m} x_i$$

contains it in its coarsest form. One now looks for a compromise; compressing the information into n centroids $(1 < n < m)$ instead of a single centroid. The question is – according to which criterion should the partitionings C_j $(j = 1, \ldots, n)$ be chosen?

If one puts (3.1.7) for each partition, by setting $\bar{x} = \bar{x}_j$ and $y = \bar{x}$ the result is

$$\sum_{i=1}^{m} \|x_i - \bar{x}\|^2 = \sum_{j=1}^{n} \sum_{i \in C_j} \|x_i - \bar{x}_j\|^2 + \sum_{j=1}^{n} m_j \|\bar{x}_j - \bar{x}\|^2. \tag{3.2.4}$$

on the basis of property (3.1.1). Since the left-hand side in (3.2.4) is a constant, the objective function

$$Z_1(C_1, \ldots, C_n) = \sum_{j=1}^{n} \sum_{i \in C_j} \|x_i - \bar{x}_j\|^2 \longrightarrow \text{minimum} \tag{3.2.5}$$

is meaningful. Therefore non-empty sets of indices $C_1, \ldots, C_n$ are sought such that the sum of the sum of the squared Euclidean distances of the cluster members from their centroids is minimised. We call this objective function the sum of squared distance criterion; it is also known as the principle of n best centroids (see Freitag 1972, MacQueen 1967), or the variance criterion (see Bock 1974).

From section 2.1 it is clear that Z_1 is invariant for $l > 1$ only with respect to orthogonal mappings and translations of the vectors x_i; therefore, for example, not with respect to scale transformations of the variables. For $l = 1$ scale invariance is guaranteed.

For partitions which are optimal with respect to Z_1, each object is nearest to the centroid of the cluster to which it is assigned, and the convex hulls of the cluster objects are disjoint (see Bock 1970, 1974, Howard 1966).

Because of (3.2.4),

$$Z_2(C_1, \dots, C_n) = \sum_{j=1}^{n} m_j \|\bar{x}_j - \bar{x}\|^2 \longrightarrow \text{maximum} \tag{3.2.6}$$

is the same as the minimisation of Z_1, that is, the maximisation of the sum of the weighted squared distances of the cluster from the centroid of the total matrix, using the numbers of members of the clusters as weights.

Since the lefthand side of (3.2.4) is constant, the objective functions

$$\frac{Z_1}{Z_2} \longrightarrow \text{Min} \qquad \text{and} \qquad \frac{Z_2}{Z_1} \longrightarrow \text{max}$$

are equivalent to (3.2.5) (see Engelman and Hartigan 1969). In the following we shall always refer to Z_1 only. Other formulations of Z_1 are to be found in (Fisher 1969) – by means of permutation matrices – and in (Cooper 1963, Duran and Odell 1974, Rao 1971, Vinod 1969) – by means of linear programming.

Weights w_i for the object vectors could have been introduced into (3.2.4) (see Fisher 1958) and hence into all the objective functions mentioned. We have not done so here, since scaling the variables is in general substantially more important than weighting of the objects.

Firstly we consider the definition of an optimal partitioning for the special case $l = 1$, since the number of the partitionings to be enumerated is drastically reduced. If the data are, in fact ordered, then

$$x_1 < x_2 < \dots < x_m, \tag{3.2.7}$$

or, if the C_j clusters are assumed to be contiguous (see Fisher 1958, Sonquist and Morgan 1964), i.e. that

$$\begin{aligned} C_j &= \{m_{j-1}+1, m_{j-1}+2, \dots, m_j\} \quad (j=1, \dots, n) \\ &\text{with } 0 = m_0 < m_1 < m_2 < \dots < m_n = m, \end{aligned} \tag{3.2.8}$$

then the number $T(m,n)$ of these contiguous partitionings of length n for m objects is

$$T(m,n) = \binom{m-1}{n-1}. \tag{3.2.9}$$

$T(m,n) \ll S(m,n)$ then holds; for example $T(m,2) = m - 1 \ll 2^{m-1} = S(m,2)$.

If the ordering of objects from the beginning has a meaning, as for example in values for time series, and it is accordingly meaningful to look for contiguous partitionings only, the condition (3.2.8) could also be assumed for $l > 1$, and only $T(m,n)$ partitionings would have to be considered.

However, let us limit ourselves in the following discussion of programs to

$l = 1$ and assume either (3.2.7) or that the order of the objects for the search of contiguous partitionings is significant.

In (Späth 1973a) an Algol procedure is presented for totally enumerating (not very effectively) the $T(m,n)$ contiguous partitionings. Since, for example,

$$T(40,8) = 15,380,937$$

$$T(50,10) = 2,054,455,634,$$

it is clear that the enumeration procedure for $m \approx 50$ has little point and is hardly capable of being carried out economically. The effectiveness of the enumeration procedure is improved significantly by using dynamic programming (see Fisher 1969, Rao 1971). Let $F_n(i)$ be the sum of the sums of squares corresponding to (3.2.5) in the partitioning of $x_1, \ldots, x_i$ ($i = 2, \ldots, m$) into n clusters. Furthermore, let

$$V_{ij} = \sum_{k=i}^{j} (x_k - \bar{x}_{i,j})^2$$

where $$\bar{x}_{i,j} = \frac{1}{j-i+1} \sum_{k=i}^{j} x_k,$$

and let

$$F_n(0) = 0 \quad (n=0, \ldots, m)$$

$$F_i(n) = 0 \quad (i=n, \ldots, m)$$

$$F_0(i) = \infty \quad (i=1, \ldots, m).$$

Then a dynamic program for minimising the objective function (3.2.5) by means of

$$F_n(j) = \min_{1 \leqq i \leqq j} [V_{ij} + F_{n-1}(i-1)]$$

$$(j=1, \ldots, m;\ n=1, \ldots, j)$$

is described in (Rao 1971), to which the reader is referred for further details.

We shall also not explain in detail the corresponding and fairly complicated subroutine ORDERD shown in Fig. U5. The main program (Fig. H4) supplies parameters as explained and contains the input and output instructions. In Fig. E4.1, the data of which is identical to that of E1.1 and E2.1, the interpretation of the line

2 43.52 1 12

for example, is that for 2 clusters the objective function has the value 43.52 and the clusters are specified by $C_1 = \{1, \ldots, 11\}$ and $C_2 = \{12, \ldots, 15\}$.

Fig. U 5

```
      SUBROUTINE ORDERD (M,X,N,Q,S)
C
C     OPTIMAL PARTITION OF ONE-DIMENSIONAL ORDERED DATA
C     X(I) (I = 1,...,M) INTO N CLUSTERS WITH MINIMAL
C     TOTAL SUM OF VARIANCES.
C
C     FOR J = 1,...,N THE VALUES OF THE INDEX I WHICH
C     MARK THE BEGINNINGS OF THE J CLUSTERS ARE STORED
C     IN THE INTEGER ARRAY Q(J,K) (K = 1,...,J).
C
C     THE ELEMENTS S(M,J) (J = 1,...,N) OF THE ARRAY S
C     CONTAIN THE MINIMAL VARIANCES BELONGING TO THE
C     J CLASSES.
C
      INTEGER   R,P,Q
C     DIMENSION X(M),  R(M,N),   Q(N,N),   S(M,N)
      DIMENSION X(100),R(100,20),Q(20,20),S(100,20)
      DO 2 J=1,N
         R(1,J)=1
         S(1,J)=0.
         DO 1 I=2,M
            S(I,J)=1.E30
    1    CONTINUE
    2 CONTINUE
      IF(N.LE.1) RETURN
      DO 5 I=2,M
         T=0.
         U=C.
         DO 4 K=1,I
            L=I-K+1
            F=X(L)
            T=T+F
            U=U+F*F
            V=U-T*T/FLOAT(K)
            P=L-1
            IF(P.EQ.0) GCTC 4
            DO 3 J=2,N
               F=S(P,J-1)+V
               IF(S(I,J).LT.F) GOTC 3
               R(I,J)=L
               S(I,J)=F
    3       CONTINUE
    4    CONTINUE
         S(I,1)=V
         R(I,1)=1
    5 CONTINUE
      DO 7 J=1,N
         K=N-J+1
         IL=M+1
         DO 6 L=1,K
            LL=K-L+1
            IU=IL-1
            IL=R(IU,LL)
            Q(K,LL)=IL
    6    CONTINUE
    7 CONTINUE
      RETURN
      END
```

A full explanation is omitted because, even with dynamic programming, a result cannot be reached within a reasonable calculation time for values of m from $m \approx 100$. Instead we must return to heuristic algorithms, to be described later, which do not guarantee that an absolute optimum will be found, but at least do produce a usable optimum in a tiny fraction of the computing time.

```
C
C     CLUSTERING OF ORDERED DATA
C
      DIMENSION X(100),P(20,20),S(100,20)
      INTEGER   P
      KI=5
      KO=6
    1 READ(KI,2) M,N
    2 FORMAT(2I5)
      IF(N.EQ.0) N=M
      IF(N.LT.0.OR.M.LE.0.OR.M.GT.100.OR.N.GT.20) STOP
      READ(KI,3) (X(I),I=1,M)
    3 FORMAT(16F5.0)
      WRITE(KO,4) M,N
    4 FORMAT('1',5X,'M =',I3,'   N=',I2)
      WRITE(KO,5)
    5 FORMAT('0')
      WRITE(KO,6) (X(I),I=1,M)
    6 FORMAT(9X,10F6.2)
      WRITE(KO,5)
      CALL ORDERD (M,X,N,P,S)
      DO 7 J=1,N
           WRITE(KO,8) J,S(M,J),(P(J,K),K=1,J)
    7 CONTINUE
    8 FORMAT(1X,I6,F9.2,2X,15I3/(18X,15I3))
      GOTO 1
      END
```

Fig. H 4

Fig. E 4.1

```
 M = 15   N=15


        0.0    0.10   0.20   0.50   2.00   2.10   2.50   3.00   3.20   4.00
        5.00   6.00   8.00  10.00  11.00


 1     175.02     1
 2      43.52     1 12
 3      16.16     1  8 13
 4       7.94     1  5 10 13
 5       3.77     1  5 10 13 14
```

```
 6    1.84   1  5  8 11 13 14
 7    1.30   1  5  8 10 11 13 14
 8    0.80   1  5  8 10 11 12 13 14
 9    0.30   1  5  8 10 11 12 13 14 15
10    0.17   1  5  7  8 10 11 12 13 14 15
11    0.05   1  4  5  7  8 10 11 12 13 14 15
12    0.03   1  4  5  7  8  9 10 11 12 13 14 15
13    0.01   1  3  4  5  7  8  9 10 11 12 13 14 15
14    0.00   1  3  4  5  6  7  8  9 10 11 12 13 14 15
15    0.0    1  2  3  4  5  6  7  8  9 10 11 12 13 14 15
```

Fig. E 4.1

A further example calculated by means of ORDERD may serve to clarify the objective criterion (3.2.5). For the case (3.1.11) when $n = 2$, ORDERD finds that the partition $C_1 = \{1, \ldots, 9\}$ and $C_2 = \{10, \ldots, 20\}$ is optimal with respect to the objective function. The criterion Z_1 thus does not find the gap in the data, unlike the criterion of maximal centroid distance mentioned in the discussion of ZWEIGO. This is, perhaps, to be expected. On the other hand, it does find it for $n = 3, \ldots, 13$ except for $n = 4$.

We now turn our attention to the more interesting general case $l \geqslant 1$, and firstly consider the relatively simple algorithm HMEANS in Fig. U6 (see Anderberg 1973, Bock 1970, Howard 1966). This starts from a given initial partitioning specified by the assignment vector p, as defined earlier, and improves this – and hence Z_1 – in a stepwise manner. In this process the cluster centroids are calculated and then each object is reassigned to the cluster whose centroid is nearest in the Euclidean sense. Computing the centroids and reassigning objects is repeated until no more changes take place or until one or more clusters become empty.

Fig. U 6

```
      SUBROUTINE HMEANS (M,L,X,P,N,S,E,D,IDR)
C
C     THE INITIAL ASSIGNMENT OF THE VECTORS X(M,*) TO N
C     CLUSTERS IS GIVEN BY THE ARRAY P, WHERE P(I) IS THE
C     CLUSTER NUMBER OF THE I-TH VECTOR,    THUS EACH P(I)
C     MUST BE SUCH THAT 1 <= P(I) <= N AND FOR EACH
C     J = 1,...,N AT LEAST ONE I WITH P(I) = J MUST EXIST,
C
C     LET D DENOTE THE SUM OF THE E(J), WHERE E(J) IS THE
C     SUM OF THE SQUARES OF THE DISTANCES BETWEEN THE
C     MEMBERS OF THE J-TH CLUSTER AND THEIR CENTROID.
```

```
C
C      EACH X(I,*) IS NOW REASSIGNED TO THE INITIAL CLUSTER
C      IT LIES CLOSEST TO.   THE ASSOCIATED CENTROIDS
C      ARE RECALCULATED AND THE PROCESS IS REPEATED.
C      D IS THUS MONOTONICALLY DECREASING.   THE ITERATION
C      IS TERMINATED IF A CLUSTER BECOMES EMPTY OR IF THREE
C      SUCCESSIVE STEPS DO NOT PRODUCE A DECREASE IN D.
C
C      THE SUBROUTINE RETURNS THE VALUES AT THE FINAL
C      CONFIGURATION FOR THE CENTROIDS S(N,L), THE
C      SUMS E(N), AND D.
C
C      IF IDR = 1, THE CURRENT VALUES OF D AND OF THE
C      VECTOR P ARE PRINTED AT EACH ITERATION.
C
C
C
       DIMENSION X(M,L),    P(M)   ,S(N,L),   E(N), Q(N)
       DIMENSION X(100,12),P(100),S(10,12),E(10),Q(10)
       INTEGER    P,Q,R
       KO=6
       ID=0
       DMAX=1.E30
       DO 1 I=1,M
             R=P(I)
             IF(R.LT.1.OR.R.GT.N) RETURN
     1 CONTINUE
     2 DO 4 J=1,N
             Q(J)=0
             DO 3 K=1,L
                   S(J,K)=0.
     3       CONTINUE
     4 CONTINUE
       DO 6 I=1,M
             R=P(I)
             Q(R)=Q(R)+1
             DO 5 K=1,L
                   S(R,K)=S(R,K)+X(I,K)
     5       CONTINUE
     6 CONTINUE
       IR=0
       DO 8 J=1,N
             E(J)=0.
             R=Q(J)
             F=0.
             IF(R.NE.0) F=1./FLOAT(R)
             IF(R.EQ.0) IR=IR+1
             DO 7 K=1,L
                   S(J,K)=S(J,K)*F
     7       CONTINUE
     8 CONTINUE
       D=0.
       DO 10 I=1,M
             R=P(I)
             F=0.
             DO 9 K=1,L
                   T=S(R,K)-X(I,K)
                   F=F+T*T
     9       CONTINUE
             E(R)=E(R)+F
             D=D+F
```

```
10 CONTINUE
   IF(IR.NE.0) RETURN
   IF(D.GE.DMAX) ID=ID+1
   IF(IDR.EQ.1) WRITE(KO,14)(ID,D,(P(J),J=1,M))
   IF(ID.GT.2) RETURN
   DMAX=D
   DO 13 I=1,M
         F=1.E30
         DO 12 J=1,N
               G=0.
               DO 11 K=1,L
                     T=S(J,K)-X(I,K)
                     G=G+T*T
11             CONTINUE
               IF(G.GE.F) GOTO 12
               F=G
               R=J
12       CONTINUE
         P(I)=R
13 CONTINUE
14 FORMAT(1X,I4,F12.2,23I3/(17X,23I3))
   GOTO 2
   END
```

Fig. U 6

The main program H5 calculates the final partition P(M) and its centroids S(N,L) for a given data matrix X(M,L) and with N1, N1 + 1, . . . ,N2 as the assumed numbers of clusters. If IP $\neq$ 0 an initial partition (P(I), I = 1, M) is read in for each N = N1, . . . ,N2 .for IP = 0 the program generates the following initial partition, which will be used as standard in later programs:

$$p_i = i - 1\,(\mathrm{mod}\, n) + 1 \quad (i = 1, \ldots, m). \tag{3.2.10}$$

Thus $p_1 = 1, p_2 = 2, \ldots, p_n = n, p_{n+1} = 1, p_{n+2} = 2, \ldots$. If n is a divisor of m, the same number of elements are assigned to each cluster at the beginning; otherwise, i.e. if $m = d{\cdot}n{+}r$ with $0 < r < n$, one more point is then initially assigned to the first r clusters than to the last $n{-}r$ clusters.

Fig. H 5

```
C
C     HMEANS-CLUSTERING WITH A GIVEN INITIAL PARTITION
C

      DIMENSION X(100,12),P(100),S(10,12),E(10)
      INTEGER P
      KI=5
      KC=6
```

```
    1 READ(KI,2) M,L,N1,N2,IDR,IP
    2 FORMAT(16I5)
      IF(M.LE.1.OR.M.GT.100.OR.L.LT.1.OR.L.GT.12.OR.
     *   N1.GT.M.OR.N1.LT.1.OR.N2.LT.N1.OR.N2.GT.10) STOP
      WRITE(KO,3) M,L,N1,N2,IDR,IP
    3 FORMAT('1','  M=',I3,'  L=',I2,'  N1=',I2,'  N2=',
     *       I2,'  IDR=',I1,'  IP=',I1)
      WRITE(KO,4)
    4 FORMAT('0')
      DO 5 K=1,L
            READ(KI,6)  (X(I,K),I=1,M)
            WRITE(KO,7) (X(I,K),I=1,M)
    5 CONTINUE
    6 FORMAT(16F5.0)
    7 FORMAT(1X,10F7.1)
      DO 13 N=N1,N2
         WRITE(KO,4)
         IF(IP.NE.0) GOTO 9
         K=0
         DO 8 I=1,M
            K=K+1
            IF(K.GT.N) K=K-N
            P(I)=K
    8    CONTINUE
         GOTO 10
    9    READ(KI,2)   (P(I),I=1,M)
   10    WRITE(KO,11) (P(I),I=1,M)
   11    FORMAT(1X,10I7)
         CALL HMEANS (M,L,X,P,N,S,E,D,IDR)
         WRITE(KO,4)
         WRITE(KO,11) (P(I),I=1,M)
         WRITE(KO,4)
         DO 12 K=1,L
            WRITE(KO,7) (S(J,K),J=1,N)
   12    CONTINUE
         WRITE(KO,4)
         WRITE(KO,7) (E(J),J=1,N),D
   13 CONTINUE
      GOTO 1
      END
```

Fig. H 5

Fig. E5.1 shows the results for a very simple example: ten points in a plane (Fig. B6) which one would intuitively divide into three clusters. The initial partitions given by (3.2.10) are printed in full for M = 2, . . . ,5 as are the resulting final partitions, the corresponding centroids, the sums of squared distances (E(J), J = 1, M), and the sum D. Optimal assignment occurs with N = 3. For N = 4 and N = 5, a disadvantage of the HMEANS algorithm appears, which users will frequently experience: instead of the desired number of clusters, a lower one is produced, namely N′ = 2 and N′ = 4, where, although the assignment is certainly not meaningless, the ‘correct’ number of clusters N = 3 is not achieved.

```
M= 10   L= 2   N1= 2   N2= 5   IDR=0   IP=0

  0.0    1.0    2.0    4.0    5.0    5.0    6.0    8.0    9.0   10.0
  1.0    2.0    0.0    8.0    7.0    9.0    7.0    4.0    3.0    5.0

    1      2      1      2      1      2      1      2      1      2

    1      1      1      2      2      2      2      2      2      2

  1.0    6.7
  1.0    6.1

  4.0   60.3   64.3

    1      2      3      1      2      3      1      2      3      1

    2      2      2      1      1      1      1      3      3      3

  5.0    1.0    9.0
  7.8    1.0    4.0

  4.8    4.0    4.0   12.7

    1      2      3      4      1      2      3      4      1      2

    3      3      3      4      4      4      4      4      4      4

  0.0    0.0    1.0    6.7
  0.0    0.0    1.0    6.1

  0.0    0.0    4.0   60.3   64.3

    1      2      3      4      5      1      2      3      4      5

    1      1      3      1      4      4      4      5      5      5

  1.7    0.0    2.0    5.3    9.0
  3.7    0.0    0.0    7.7    4.0

 37.3    0.0    0.0    3.3    4.0   44.7
```

Fig. E 5.1

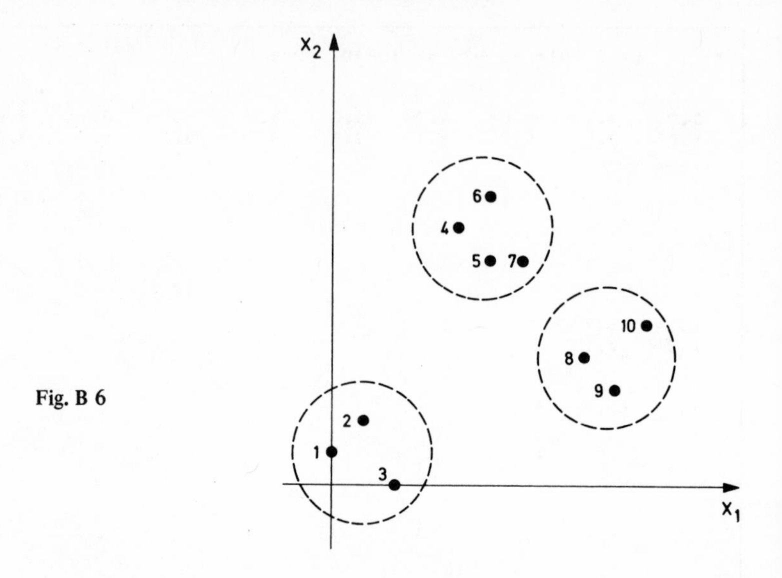

Fig. B 6

Like all non-enumerative algorithms, HMEANS does not necessarily find an optimal partition. It only guarantees that Z_1 does not become greater at each step – in the standard case the first steps decrease it. Of course, the result always depends on the initial partition selected. In the next algorithm, for which LEADER and JOINER as well as HMEANS may serve as preliminaries, to compute a satisfactory initial partitioning, we shall go into this question more thoroughly.

In order to be able to describe the next algorithm, KMEANS, which theoretically and certainly for purposes of application is the most important one here, we require several more formulae to show how the present centroids and sums of squared distances can be simply expressed, when transferring one or more points from one cluster to another, in terms of the values available for them and of the points being relocated.

Let C_j denote a cluster with a number of elements $m_j \geqslant 2$. Let C_k be a non-empty but proper subset contained in C_j, i.e. $C_k \neq \phi$, $C_k \neq C_j$, $C_k \subset C_j$. Finally let $C_p = C_j - C_k$ be the (non-empty) difference set between C_j and C_k. Given

$$\bar{x}_q = \frac{1}{m_q} \sum_{i \in C_q} x_i \quad (q = j, k, p) \tag{3.2.11}$$

as defining the centroids and with

$$e_q = \sum_{i \in C_q} \|x_i - \bar{x}_q\|^2 \quad (q = j, k, p) \tag{3.2.12}$$

the sum of the squared distances of the elements of the subset from its centroid can be calculated. By squaring and using (3.1.6), e_q can be written

$$e_q = \sum_{i \in C_q} \|x_i\|^2 - m_q \|\bar{x}_q\|^2. \tag{3.2.13}$$

The centroid x_p and the sum of the squared distances e_p can be represented as functions of $\bar{x}_j$, $\bar{x}_k$, e_j and e_k as follows (see Howard 1966)

$$\bar{x}_p = \frac{m_j \bar{x}_j - m_k \bar{x}_k}{m_j - m_k} \tag{3.2.14}$$

$$e_p = e_j - e_k - \frac{m_j m_k}{m_j - m_k} \|\bar{x}_j - \bar{x}_k\|^2. \tag{3.2.15}$$

The proof of (3.2.15) is:

$$\begin{aligned}
e_p &= \sum_{i \in C_p} \|x_i\|^2 - m_p \|\bar{x}_p\|^2 \\
&= \sum_{i \in C_j} \|x_i\|^2 - \sum_{i \in C_k} \|x_i\|^2 - \frac{1}{m_p} \|m_j \bar{x}_j - m_k \bar{x}_k\|^2 \\
&= (e_j + m_j \|\bar{x}_j\|^2) - (e_k + m_k \|\bar{x}_k\|^2) \\
&\quad - \frac{1}{m_j - m_k} \|m_j \bar{x}_j - m_k \bar{x}_k\|^2 \\
&= e_j - e_k + \left(m_j - \frac{m_j^2}{m_j - m_k}\right) \|\bar{x}_j\|^2 \\
&\quad - \left(m_k + \frac{m_k^2}{m_j - m_k}\right) \|\bar{x}_k\|^2 + \frac{2\, m_j m_k}{m_j - m_k} \bar{x}_j^T \bar{x}_k \\
&= e_j - e_k - \frac{m_j m_k}{m_j - m_k} \|\bar{x}_j - \bar{x}_k\|^2.
\end{aligned}$$

For the special case, which will be required later, that $C_k = \{k\}$ and is therefore composed of a single element, (3.2.14) and (3.2.15) become

$$\bar{x}_p = \frac{m_j \bar{x}_j - x_k}{m_j - 1} \tag{3.2.16}$$

$$e_p = e_j - \frac{m_j}{m_j - 1} \|\bar{x}_j - x_k\|^2. \tag{3.2.17}$$

Should C_p be produced not by division, as previously, but by uniting, that is $C_p = C_j \cup C_k$ with $C_j \cap C_k = \phi$, the corresponding formulae are

$$\bar{x}_p = \frac{m_j \bar{x}_j + m_k \bar{x}_k}{m_j + m_k} \tag{3.2.18}$$

$$e_p = e_j + e_k + \frac{m_j m_k}{m_j + m_k} \|\bar{x}_j - \bar{x}_k\|^2 \tag{3.2.19}$$

and, for $C_k = \{k\}$

$$\bar{x}_p = \frac{m_j \bar{x}_j + x_k}{m_j + 1} \tag{3.2.20}$$

$$e_p = e_j + \frac{m_j}{m_j + 1} \|\bar{x}_j - x_k\|^2. \tag{3.2.21}$$

as can be verified in a similar way.

In the algorithm KMEANS in Fig. U7 only the special cases (3.2.16), (3.2.17), (3.2.20) and (3.2.21) are used. Starting from a given partition, each point x_i in turn is transferred experimentally from its cluster $p_i = r$ into every other cluster $j = 1, \ldots, n$ $(j \neq r)$. If for at least one $j \neq r$,

$$\frac{m_r}{m_r - 1} \|\bar{x}_r - x_i\|^2 > \frac{m_j}{m_j + 1} \|\bar{x}_j - x_i\|^2 \qquad (p_i = r;\ j \neq r), \tag{3.2.22}$$

then x_i is allocated to the cluster v for which the right-hand side of (3.2.22) is minimised, which means that $e_r + e_v$ becomes

$$e_r - \frac{m_r}{m_r - 1} \|\bar{x}_r - x_i\|^2 + e_v + \frac{m_v}{m_v + 1} \|\bar{x}_v - x_i\|^2 \tag{3.2.23}$$

and accordingly the total sum $d = \sum_{j=1}^{n} e_j$ is reduced as much as possible. If this is not the case, the next point is considered. As many passes are performed through the objects $i = 1, \ldots, m$ as are needed, until there are no more changes.

Fig. U 7

```
      SUBROUTINE KMEANS (M,L,X,P,N,S,E,D,IDR)
C
C     THE INITIAL ASSIGNMENT OF THE VECTORS X(M,L) TO N CLUSTERS
C     IS GIVEN BY THE ARRAY P, WHERE P(I) IS THE CLUSTER NUMBER
C     OF THE I-TH VECTOR.    THUS EACH P(I) MUST BE SUCH THAT
C     1 <= P(I) <= N AND FOR EACH J = 1,...,N AT LEAST ONE I
C     WITH P(I) = J MUST EXIST.
```

```
C
C     LET D DENOTE THE SUM OF THE E(J), WHERE E(J) IS THE SUM
C     OF THE SQUARES OF THE DISTANCES BETWEEN THE MEMBERS OF
C     THE J-TH CLUSTER AND THEIR CENTROID.
C
C     THE SUBROUTINE MINIMISES D AS FAR AS POSSIBLE BY
C     REPEATED EXCHANGES OF CLUSTER MEMBERS.   THE P(I)
C     ARE CORRESPONDINGLY MODIFIED WITHOUT, HOWEVER,
C     CHANGING THE NUMBER OF CLUSTERS.
C
C     THE SUBROUTINE RETURNS THE VALUES AT THE FINAL
C     CONFIGURATION FOR THE CENTROIDS S(N,L), THE
C     SUMS E(N), AND D.
C
C     IF IDR = 1, THE CURRENT VALUES OF D AND OF THE
C     VECTOR P ARE PRINTED AT EACH ITERATION.
C
C     DIMENSION X(M,L),    P(M)   ,S(N,L),   E(N),  Q(N)
      DIMENSION X(100,12),P(100),S(10,12),E(10),Q(10)
      INTEGER   P,Q,R,U,V,W
      KO=6
      DO 2 J=1,N
            Q(J)=0
            E(J)=0.
            DO 1 K=1,L
                  S(J,K)=0.
    1       CONTINUE
    2 CONTINUE
      DO 4 I=1,M
            R=P(I)
            IF(R.LT.1.OR.R.GT.N) RETURN
            Q(R)=Q(R)+1
            DO 3 K=1,L
                  S(R,K)=S(R,K)+X(I,K)
    3       CONTINUE
    4 CONTINUE
      DO 6 J=1,N
            R=Q(J)
            IF(R.EQ.0) RETURN
            F=1./FLOAT(R)
            DO 5 K=1,L
                  S(J,K)=S(J,K)*F
    5       CONTINUE
    6 CONTINUE
      DO 8 I=1,M
            R=P(I)
            F=0.
            DO 7 K=1,L
                  T=S(R,K)-X(I,K)
                  F=F+T*T
    7       CONTINUE
            E(R)=E(R)+F
    8 CONTINUE
      D=0.
      DO 9 J=1,N
            D=D+E(J)
    9 CONTINUE
      I=0
      IT=0
   10 I=I+1
      IF(I.GT.M) I=I-M
      IF(IT.EQ.M) RETURN
```

```
      R=P(I)
      U=Q(R)
      IF(U.LE.1) GOTO 10
      H=FLOAT(U)
      H=H/(H-1.)
      F=0.
      DO 11 K=1,L
            T=S(R,K)-X(I,K)
            F=F+T*T
   11 CONTINUE
      A=H*F
      B=1.E30
      DO 13 J=1,N
            IF(J.EQ.R) GOTO 13
            U=Q(J)
            H=FLOAT(U)
            H=H/(H+1.)
            F=0.
            DO 12 K=1,L
                  T=S(J,K)-X(I,K)
                  F=F+T*T
   12       CONTINUE
            F=H*F
            IF(F.GT.B) GOTO 13
            B=F
            V=J
            W=U
   13 CONTINUE
      IF(B.LT.A) GOTO 14
      IT=IT+1
      GOTO 10
   14 IT=0
      E(R)=E(R)-A
      E(V)=E(V)+B
      D=D-A+B
      H=FLOAT(Q(R))
      G=FLOAT(W)
      A=1./(H-1.)
      B=1./(G+1.)
      DO 15 K=1,L
            F=X(I,K)
            S(R,K)=(H*S(R,K)-F)*A
            S(V,K)=(G*S(V,K)+F)*B
   15 CONTINUE
      P(I)=V
      Q(R)=Q(R)-1
      Q(V)=Q(V)+1
      IF(IDR.EQ.1) WRITE(KO,16) I,D,(P(U),U=1,M)
   16 FORMAT(1X,I4,F12.2,23I3/(17X,23I3))
      GOTO 10
      END
```

Fig. U 7

The essential difference from HMEANS, where new centroids are computed only after all objects have been allocated, is that here the centroids are modified each time one object is transferred. The results generally depend on the sequence in which the objects are considered.

Since with clusters of one element the algorithm no longer functions in the form described, any x which forms a cluster on its own is not taken into account. This often prevents very good or fully optimal partitions from being found, when $m/n < 5$; if m/n is large enough, this problem does not normally appear, and for practical purposes there is no limitation to the application of the algorithm. In contrast to HMEANS, the number of clusters cannot be reduced.

Following the principle of the optimal exchange of the points x_i taken in turn, which we shall apply to further objective functions in this chapter and will call the 'KMEANS Principle' for short, the KMEANS algorithm is stepwise optimal. A local optimum in some sense, though not, however, a global one is found. The attainment of a global optimum could only be proved by enumeration of all $S(m,n)$ partitions, which, though it could be done in principle, is not practicable. It is recommended to use several initial partitions and to use those final partitions with minimal value of the objective function.

With the aid of the following examples, we shall report on the benefits of the local optima which have been attained, and with other empirically-gained knowledge. These applications are very numerous and detailed at this stage, since we consider KMEANS to be the most important cluster algorithm for metric data matrices. This is despite problems of scaling, which we should be able to deal with in section 4.4, by using another, but essentially more complicated, objective function (See also Späth 1977a, 1977b)

KMEANS Cluster Algorithm

Fig. B 7

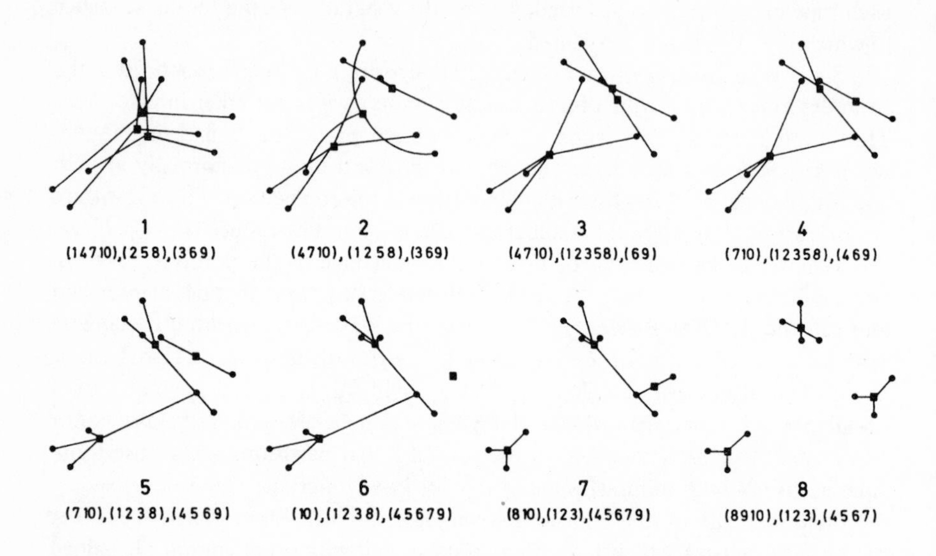

Firstly, Fig. B7 shows the individual steps of the algorithm for the configuration of points already given in Fig. B6, with $n = 3$ and beginning with the standard initial partition (3.2.10). At each step C_1, C_2 and C_3 are given (the serial numbers of the points refer to B6), and in addition the allocation is shown graphically by linking the points to the cluster centroids. The results were obtained using the main program H6, which is almost identical with H5, with IDR = 1; Fig. E6.1 gives the results for the same example with IDR = 0 and N = 2, . . . ,5, and is directly comparable with Fig. E5.1.

Fig. H 6

```
C
C     KMEANS-CLUSTERING WITH A GIVEN INITIAL PARTITION
C
      DIMENSION X(100,12),P(100),S(10,12),E(10)
      INTEGER P
      KI=5
      KO=6
    1 READ(KI,2) M,L,N1,N2,IDR,IP
    2 FORMAT(16I5)
      IF(M.LE.1.OR.M.GT.100.OR.L.LT.1.OR.L.GT.12.OR.
     *   N1.GT.M.OR.N1.LT.1.OR.N2.LT.N1.OR.N2.GT.10) STOP
      WRITE(KO,3) M,L,N1,N2,IDR,IP
```

```
  3 FORMAT('1','   M=',I3,'   L=',I2,'   N1=',I2,'   N2=',
   *          I2,'  IDR=',I1,'   IP=',I1)
    WRITE(KO,4)
  4 FORMAT('0')
    DO 5 K=1,L
         READ(KI,6)   (X(I,K),I=1,M)
         WRITE(KO,7)  (X(I,K),I=1,M)
  5 CONTINUE
  6 FORMAT(16F5.0)
  7 FORMAT(1X,10F7.1)
    DO 13 N=N1,N2
          WRITE(KO,4)
          IF(IP.NE.0) GOTO 9
          K=0
          DO 8 I=1,M
               K=K+1
               IF(K.GT.N) K=K-N
               P(I)=K
  8       CONTINUE
          GOTO 10
  9       READ(KI,2)     (P(I),I=1,M)
 10       WRITE(KO,11)   (P(I),I=1,M)
 11       FORMAT(1X,10I7)
          CALL KMEANS (M,L,X,P,N,S,E,D,IDR)
          WRITE(KO,4)
          WRITE(KO,11)  (P(I),I=1,M)
          WRITE(KO,4)
          DO 12 K=1,L
                WRITE(KO,7) (S(J,K),J=1,N)
 12       CONTINUE
          WRITE(KO,4)
          WRITE(KO,7)  (E(J),J=1,N),D
 13 CONTINUE
    GOTO 1
    END
```

Fig. E 6.1

```
M= 10   L= 2   N1= 2   N2= 5   IDR=0   IP=0

   0.0    1.0    2.0    4.0    5.0    5.0    6.0    8.0    9.0   10.0
   1.0    2.0    0.0    8.0    7.0    9.0    7.0    4.0    3.0    5.0

     1      2      1      2      1      2      1      2      1      2

     1      1      1      2      2      2      2      2      2      2

   1.0    6.7
   1.0    6.1
```

4.0	60.3	64.3							
1	2	3	1	2	3	1	2	3	1
2	2	2	3	3	3	3	1	1	1
9.0	1.0	5.0							
4.0	1.0	7.7							
4.0	4.0	4.8	12.7						
1	2	3	4	1	2	3	4	1	2
3	3	3	2	2	2	2	1	1	4
8.5	5.0	1.0	10.0						
3.5	7.7	1.0	5.0						
1.0	4.7	4.0	0.0	9.7					
1	2	3	4	5	1	2	3	4	5
2	2	2	1	1	1	1	3	4	5
5.0	1.0	8.0	9.0	10.0					
7.7	1.0	4.0	3.0	5.0					
4.7	4.0	0.0	0.0	0.0	8.7				

Fig. E 6.1

In the second example we have again taken the Cartesian co-ordinates of the 22 towns from Fig. T1, which were used in B2 and B3. The results for $n = 3,4,5$ starting with the initial partition (3.2.10) are given in Fig. B8, in which the towns belonging to one cluster are marked off from the others by being ringed by broken lines and by broken-and-dotted lines. By using another initial allocation, namely

$$\begin{aligned} p_i &= j \;\; (j=1,\dots,n-1;\; i=(j-1)\,r+1,\dots,j\,r) \\ p_i &= n \;\; \left(i=(n-1)\,r+1,\dots,m;\; r=\left[\frac{m}{n}\right]\right) \end{aligned} \tag{3.2.24}$$

we obtain the same final partitioning. Fig. T2 shows the two initial allocations for $n = 2, \dots ,5$, the values (identical) for $Z_1 = d$, as well as the number of passes, and the number of points allocated to another cluster at each pass.

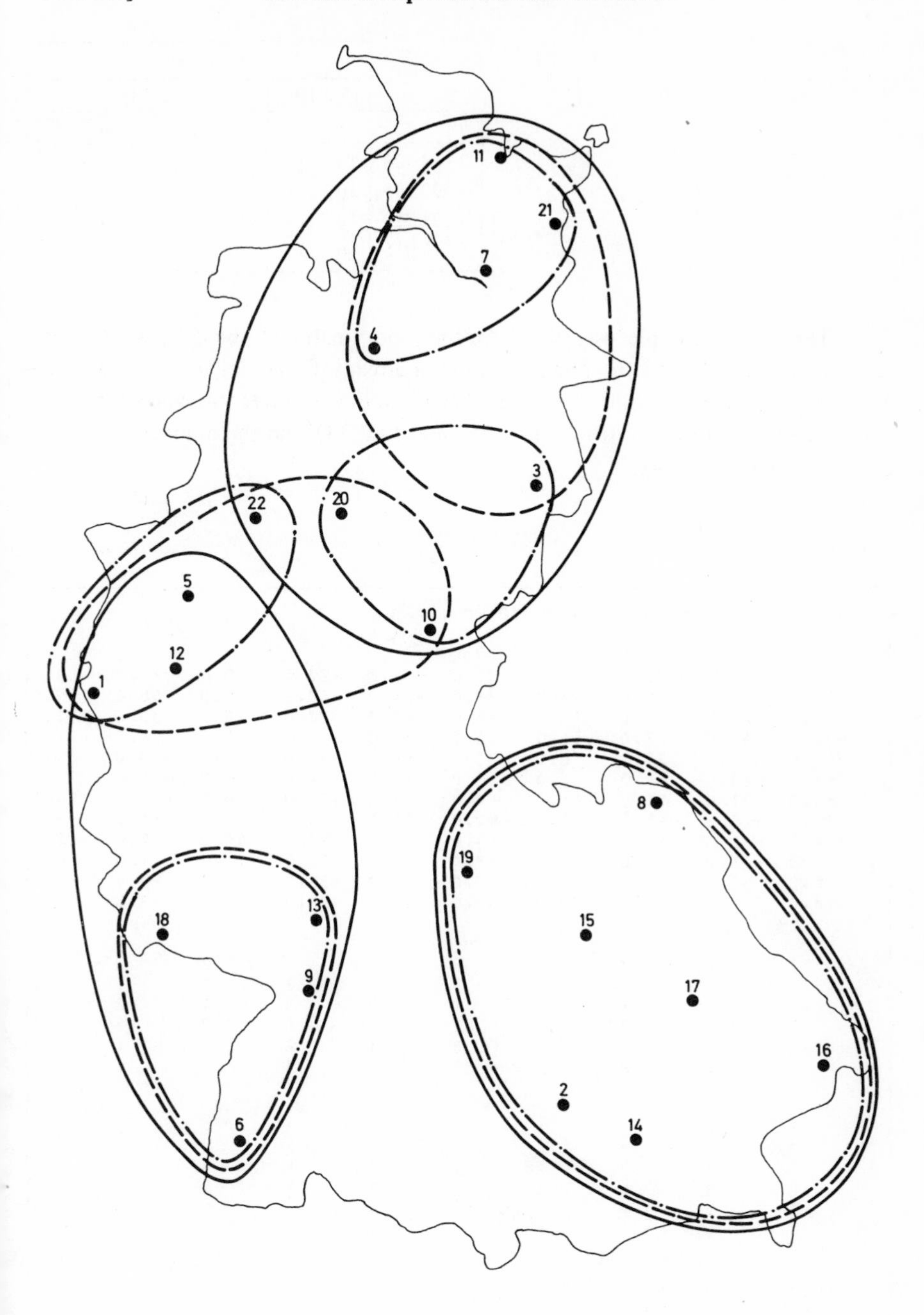

Fig. B 8

Table T2

		Passes at						
		(3.2.10)		(3.2.24)				
n	d	1	2	1	2	3	4	5
2	64 408	13	3	7				
3	39 399	10		13	3	3	2	1
4	21 719	12	1	15	6	2	5	
5	17 015	16	5	14	4			

The third example uses the Cartesian coordinates of the 59 towns given earlier in Fig. B5, and the final partitioning already found for $n = 3$ as well as the centroids and the sums of squared distances. Fig. T3 gives the values obtained for d with the initial partitions (3.2.10) and (3.2.24), and the number of points transferred at each pass.

Fig. E 6.2

```
 M= 59   L= 2   N1= 3   N2=10   IDR=0   IP=0


  54.0    0.0  -31.0    8.0    1.0  -36.0  -22.0    0.0   34.0   28.0
  12.0  -21.0   -6.0   21.0   38.0  -24.0  -38.0   86.0   58.0   -9.0
  70.0  -20.0  -43.0   59.0   -5.0   83.0   27.0   12.0   30.0   31.0
 -57.0   44.0    7.0   54.0   65.0  -35.0   46.0    5.0   56.0  -21.0
 -40.0  -43.0   57.0    0.0   25.0   56.0  -34.0  -24.0  -25.0   64.0
  63.0   37.0   -5.0    2.0  -18.0  -10.0   12.0  -40.0  -16.0
 -65.0   71.0   53.0  111.0   -9.0   52.0  -76.0   20.0  129.0   84.0
 -38.0  -26.0  -41.0   45.0  -90.0   10.0   35.0  -57.0   -1.0   -3.0
 -74.0   70.0   44.0  -26.0  114.0  -41.0  153.0  -49.0  -65.0  -12.0
  28.0  -28.0   -7.0   -8.0   -8.0   25.0   79.0  118.0    4.0   54.0
  45.0   51.0  -21.0    0.0   15.0  -25.0   56.0   36.0   49.0  -26.0
 -48.0  155.0  -24.0   28.0  -58.0   82.0  -58.0  -28.0   28.0


     1      2      3      1      2      3      1      2      3      1
     2      3      1      2      3      1      2      3      1      2
     3      1      2      3      1      2      3      1      2      3
     1      2      3      1      2      3      1      2      3      1
     2      3      1      2      3      1      2      3      1      2
     3      1      2      3      1      2      3      1      2


     3      1      2      1      2      2      3      2      1      1
     3      2      3      2      3      2      2      3      3      2
     3      2      2      3      1      3      1      3      3      3
     2      3      2      3      3      2      1      1      3      2
     2      2      3      2      2      3      2      2      2      3
     3      1      3      2      3      1      3      2      2


  17.0  -20.0   39.7
 109.6   26.8  -39.0


11734.527970.337303.677006.6
```

		(3.2.10)								(3.2.24)					
n	d	1	2	3	4	5	6	7	d	1	2	3	4	5	6
3	77 007	36	10	2					90 931	37	10	7	1		
4	49 597	47	17	10	3				49 597	36	11	2			
5	39 509	42	14	4	1				39 451	38	15	2	1		
6	30 533	48	15	10	8	1			32 529	47	20	5	9	4	1
7	24 431	44	20	7	4	3	3	3	32 541	44	9				
8	21 883	45	18	4	1				23 971	45	14				
9	19 129	54	16	6	6	5	5	1	20 236	49	15	6	4	1	1
10	17 649	48	18	7	3	2			16 862	52	16	5	5	2	1

Table T3

In Figs. T2 and T3 it can be seen that the initial allocation (3.2.10) is generally more favourable with respect to the value obtained for d than is (3.2.24), which may be caused by the fact that the number of points per cluster is more equally distributed in (3.2.10) than in (3.2.24). Furthermore, it can be seen – and this is the experience of many other people too – that the task is usually completed in 6 to 7 passes. Usually the number of transferred points goes down with increasing pass number. It is almost always the case that, the more passes that are carried out, the smaller the value of the objective function. However, the differences in the objective function due to various initial allocations are not very great in terms of percentages. It is often more useful to apply KMEANS directly using a number of random initial partitions, rather than to employ a relatively cumbersome process such as JOINER or HMEANS. This is because their final partition allocations are not certain to lead to an absolute (global) optimum as a result of the later use of KMEANS.

In the next example we consider 23 (artificial) seasonal curves $x_{i.}^T$, which are characterised by values at 12 points of time (e.g. months). The curves are presented in Fig. B9 and their coordinates are given in Fig. E6.3. The scaling was chosen so that the following relation holds:

$$\sum_{k=1}^{12} x_{ik} = 100 \quad (i=1,\ldots,23), \tag{3.2.25}$$

and thus the curves all have the same L_1 – norm. The problem is to find classes of similar curves with the help of cluster analysis (Späth 1978) indicates how classification of roughly 30,000 seasonal curves was successfully used to set up a basis for a forecasting system. The principle can be applied to time series of all kinds, whether for monthly returns, number of bookings for foreign travel, the number of customers per hour asking for certain types of commodities at pre-arranged checkouts in a store, or time series from economics as in (Elton and Gruber 1971). One interesting application in medicine is the classification of illnesses according to the leisure pursuits of patients in hospital, described in (Lades and Schläger 1975). The method based on the L_1 – metric given in section 3.5 should be even more adequate.

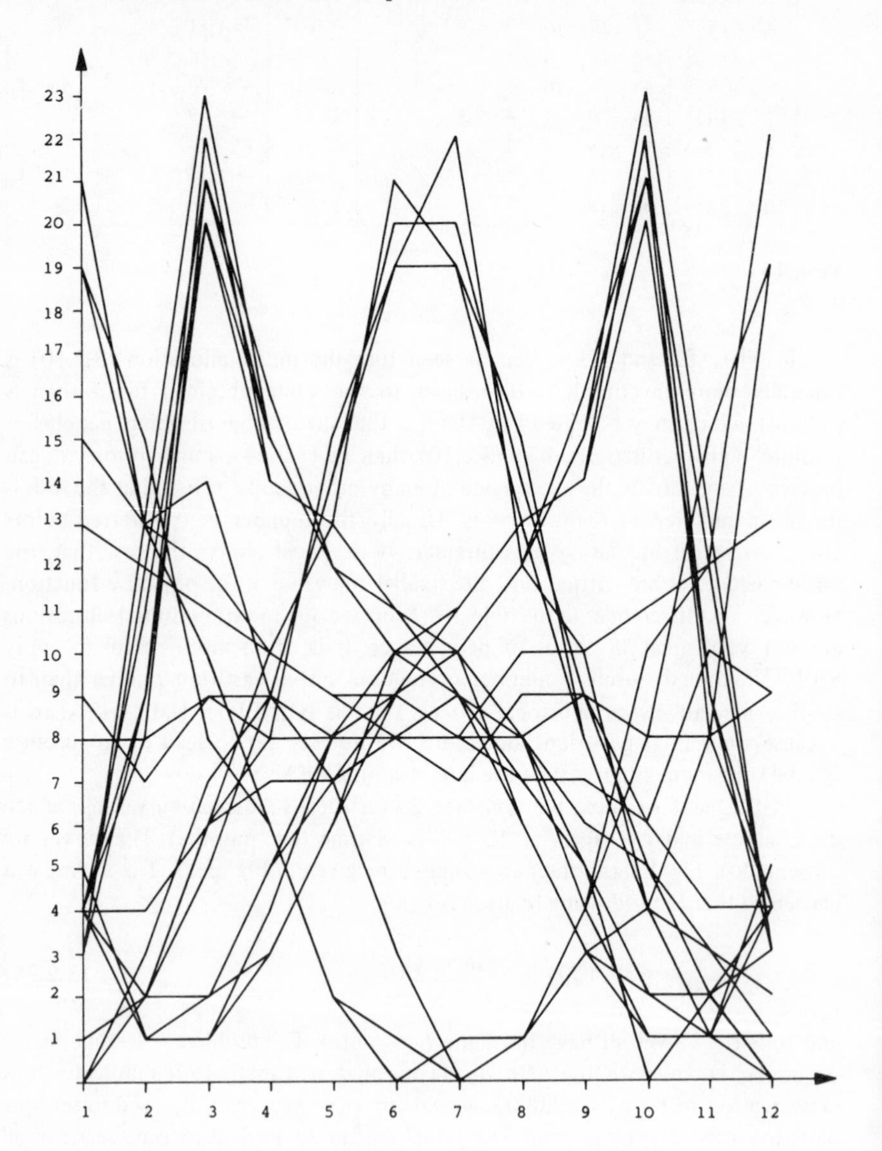

Fig. B 9

```
M= 23   L=12   N1= 2   N2= 9   IDR=0   IP=0

   3.0    3.0    3.0    4.0    4.0    4.0    0.0    1.0    0.0    0.0
   4.0    5.0    4.0    4.0    4.0   21.0   19.0   19.0    8.0    9.0
  13.0   15.0    4.0
   9.0   12.0   13.0   11.0    9.0    8.0    2.0    2.0    2.0    3.0
   2.0    1.0    1.0    1.0    1.0   13.0   15.0   13.0    8.0    7.0
  12.0   13.0    4.0
  22.0   21.0   23.0   20.0   20.0   21.0    5.0    6.0    6.0    7.0
   2.0    2.0    1.0    2.0    1.0   10.0    8.0   14.0    9.0    9.0
  11.0   12.0    6.0
  14.0   16.0   15.0   15.0   14.0   15.0    9.0    8.0   10.0    9.0
   5.0    5.0    3.0    3.0    4.0    6.0    6.0    9.0    8.0    9.0
  10.0   10.0    7.0
  13.0   12.0   12.0   12.0   13.0   12.0   14.0   15.0   13.0   14.0
   8.0    7.0    6.0    6.0    8.0    2.0    2.0    7.0    8.0    8.0
   9.0    9.0    7.0
  11.0   10.0   10.0   11.0   11.0   10.0   20.0   19.0   21.0   19.0
   9.0    9.0    9.0   10.0   10.0    0.0    1.0    3.0    9.0    8.0
   9.0    8.0    8.0
  10.0   10.0    9.0    9.0    9.0    9.0   20.0   22.0   19.0   19.0
  10.0   11.0   10.0   10.0   11.0    1.0    0.0    0.0    8.0    7.0
   8.0    9.0    8.0
   8.0    6.0    6.0    8.0    7.0    8.0   14.0   12.0   15.0   13.0
  12.0   13.0   12.0   12.0   13.0    1.0    1.0    1.0    8.0    9.0
   7.0    8.0   10.0
   4.0    3.0    3.0    5.0    5.0    5.0    9.0   10.0    8.0    8.0
  15.0   14.0   15.0   16.0   14.0    4.0    4.0    3.0    9.0    9.0
   7.0    6.0   10.0
   1.0    2.0    1.0    0.0    2.0    2.0    5.0    4.0    4.0    4.0
  21.0   20.0   23.0   21.0   22.0    8.0   10.0    4.0    8.0    6.0
   6.0    5.0   11.0
   1.0    1.0    1.0    2.0    1.0    2.0    2.0    1.0    1.0    3.0
   8.0    9.0   13.0   12.0    9.0   15.0   12.0    8.0    8.0   10.0
   4.0    3.0   12.0
   4.0    4.0    4.0    3.0    5.0    4.0    0.0    0.0    1.0    1.0
   4.0    4.0    3.0    3.0    3.0   19.0   22.0   19.0    9.0    9.0
   4.0    2.0   13.0

     1      2      1      2      1      2      1      2      1      2
     1      2      1      2      1      2      1      2      1      2
     1      2      1

     2      2      2      2      2      2      2      2      2      2
     1      1      1      1      1      1      1      1      1      1
     2      2      1

   9.2    4.2
   6.0    8.0
   5.8   14.5
   5.9   12.1
   6.3   12.3
   6.9   13.2
   6.9   12.7
   8.4    9.3
  10.3    6.1
  14.0    3.0
  10.5    1.8
   9.8    2.7

3039.3 2052.7 5091.9
```

Fig. E 6.3

However, back to our special example. Fig. E6.3 shows the allocations for $n = 2$ produced, as usual, by using H6, together with the two centroids and the sums of the squared distances. The centroids obtained are also given in graphical form in Fig. B10. Comparison with the seasonal curves in B9 shows that the value selected for the number of clusters is certainly too low. For that reason we also give in Figs. B11 to B15 the centroids obtained with (3.2.10) for $n = 3, \ldots, 7$. The characteristic peaks from B9 do not reappear in simplified form until B13 for $n = 5$, so in this case we might recommend a classification based on this number of clusters. This is supported by consideration of the values of the objective function for the various numbers of clusters, in Table T4:

Table T4

n	2	3	4	5	6	7
d	5 092	2 992	1 323	563	298	239

The reduction in d between $n = 4$ and $n = 5$ is not for later values equalled either in size or gradient.

When using a table like T4, before deciding what is a reasonable value for the number of clusters – even having used complicated statistical significance criteria – it is necessary to take care, since one is not comparing optimal values of the objective function, but the values of local optima of possibly different merit. For this reason, any inference obtained in such a way can do no more than support the choice of a specific value for the number of clusters which leads to a particularly good and reasonable interpretation.

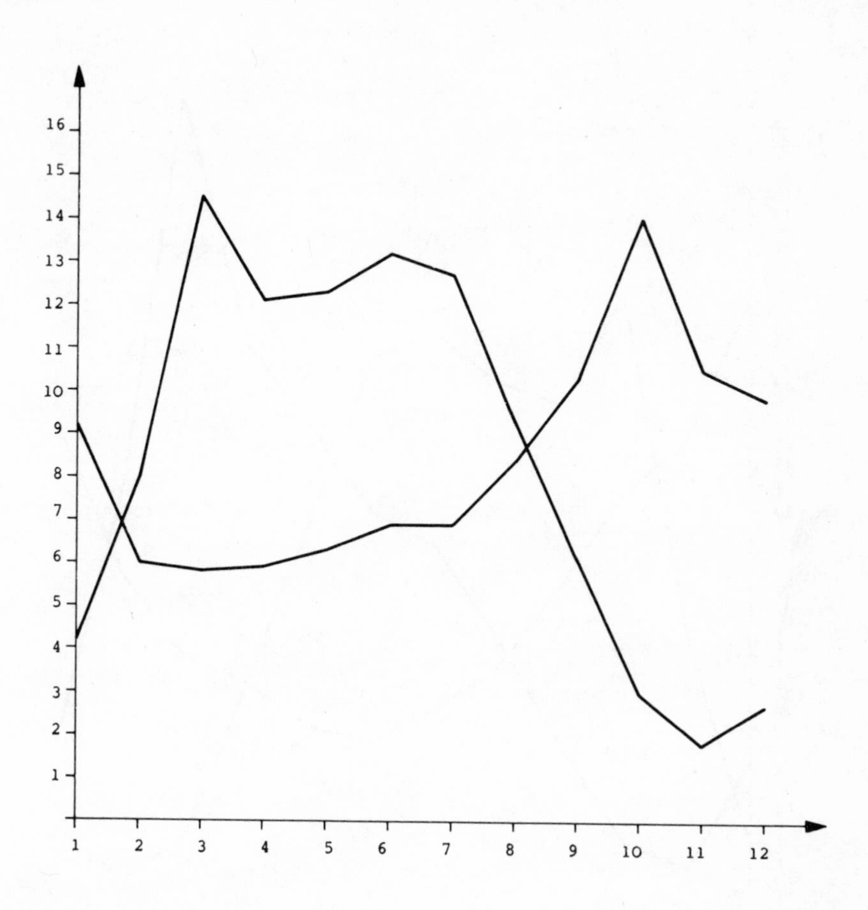

Fig. B 10

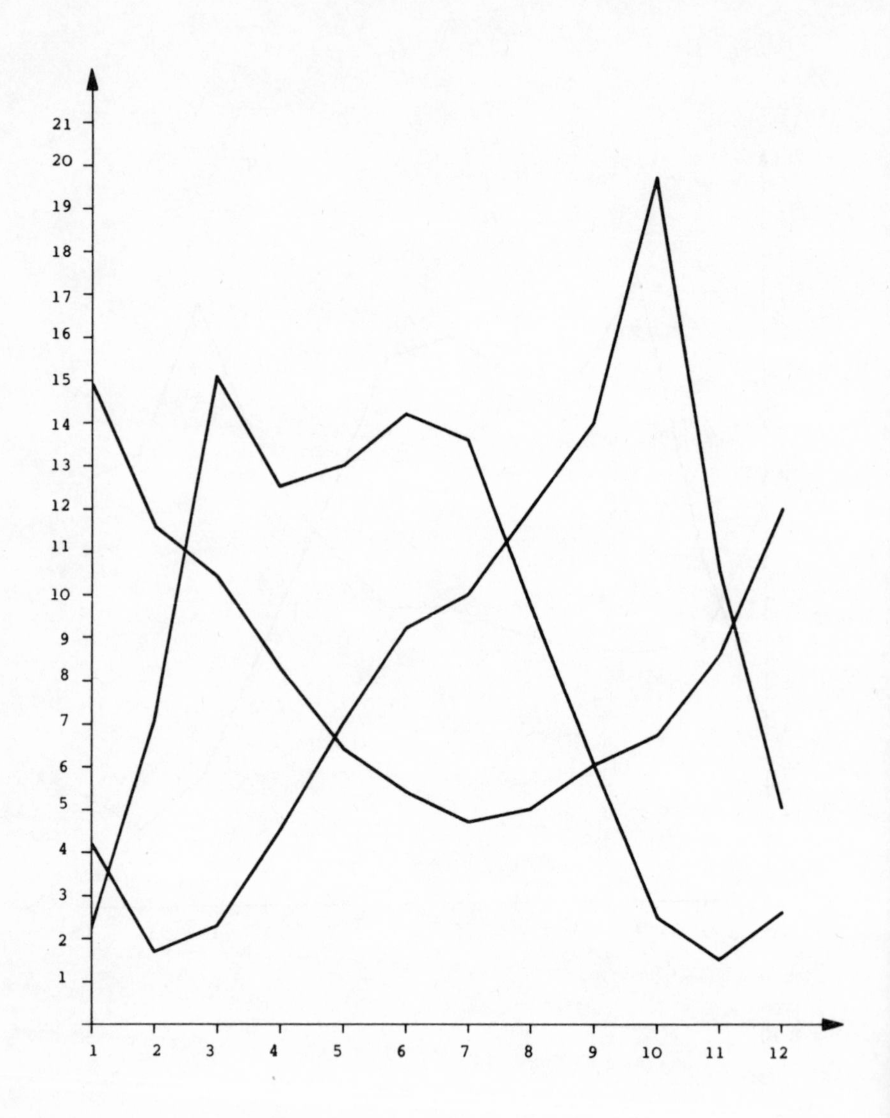

Fig. B 11

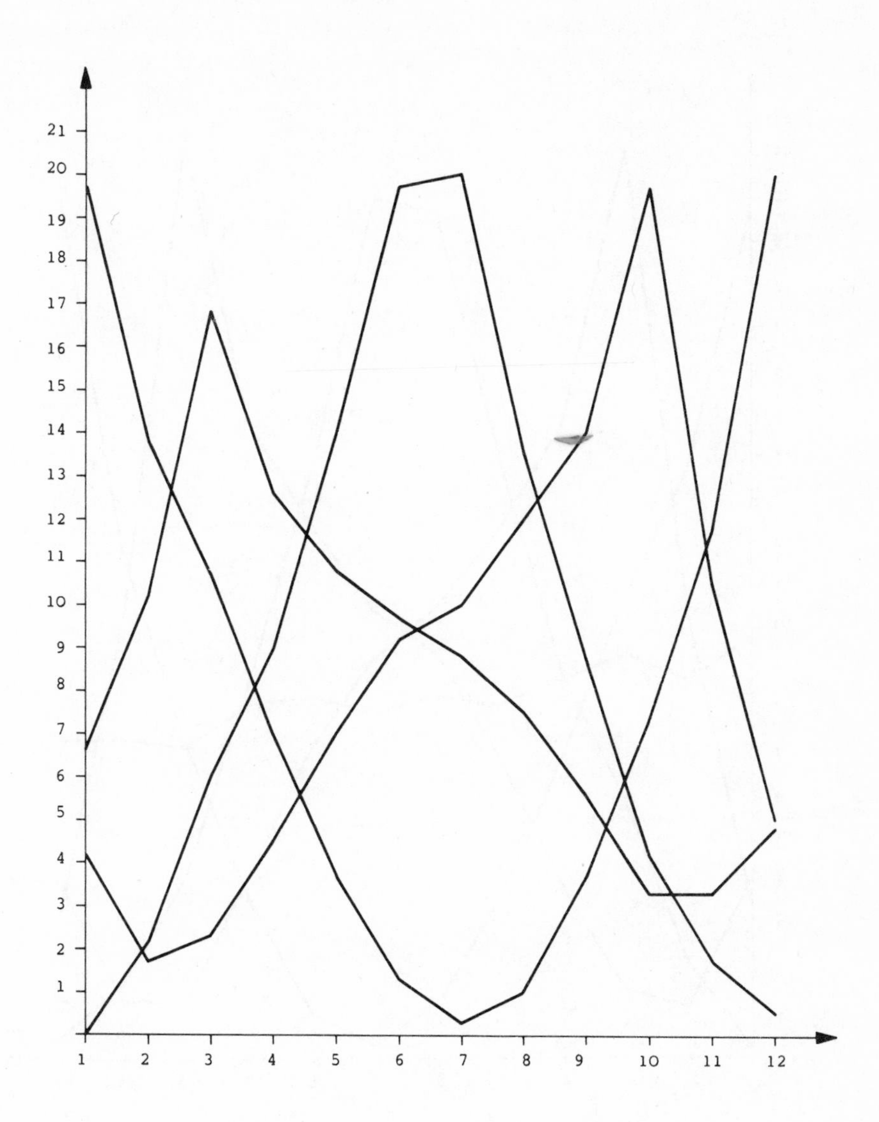

Fig. B 12

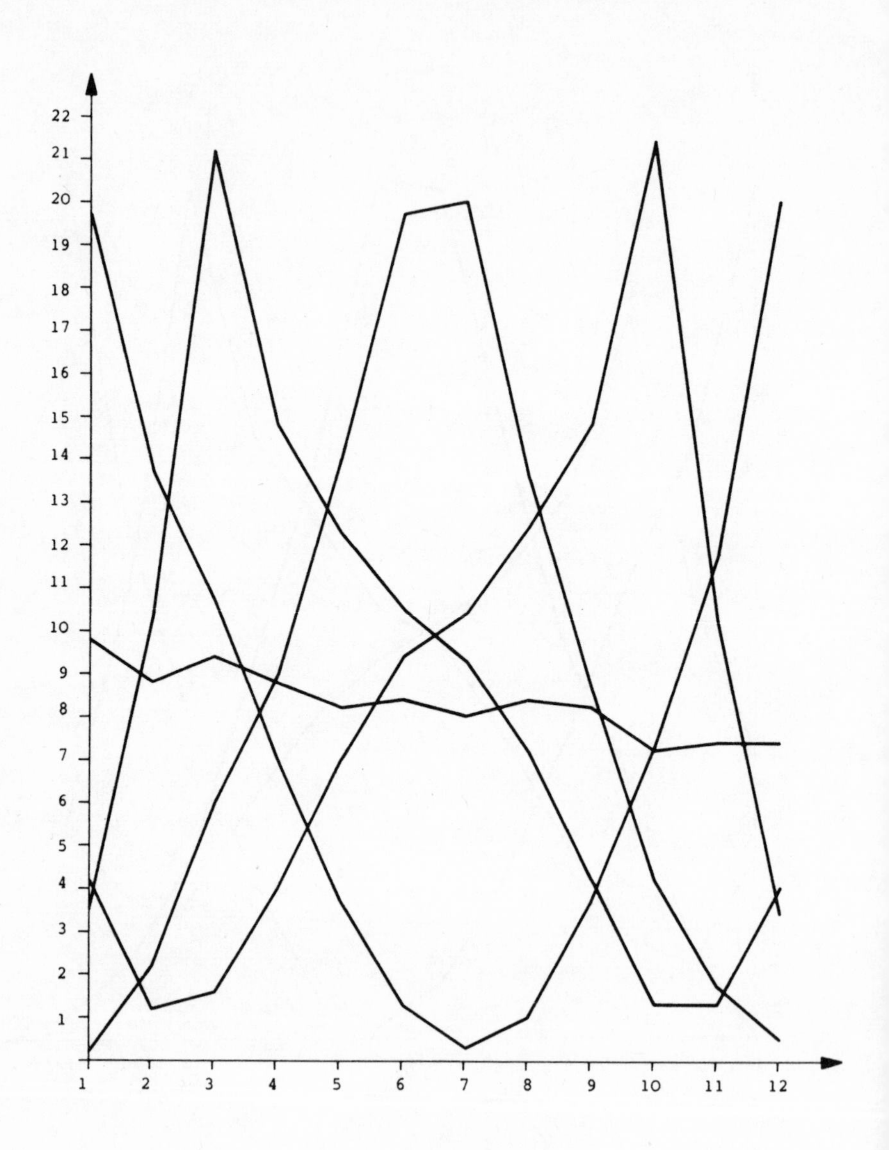
22
21
20
19
18
17
16
15
14
13
12
11
10
9
8
7
6
5
4
3
2
1
1
2
3
4
5
6
7
8
9
10
11
12

Fig. B 13

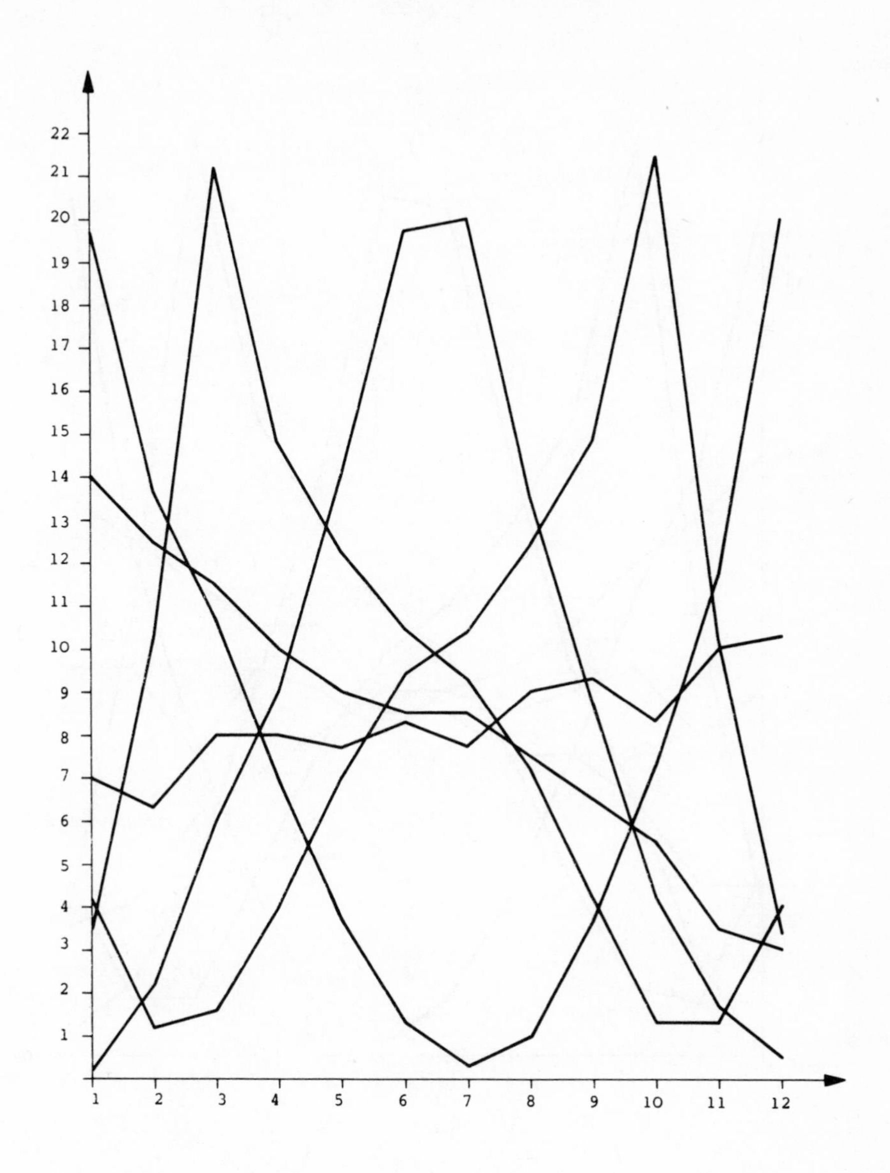
22
21
20
19
18
17
16
15
14
13
12
11
10
9
8
7
6
5
4
3
2
1
1
2
3
4
5
6
7
8
9
10
11
12

Fig. B 14

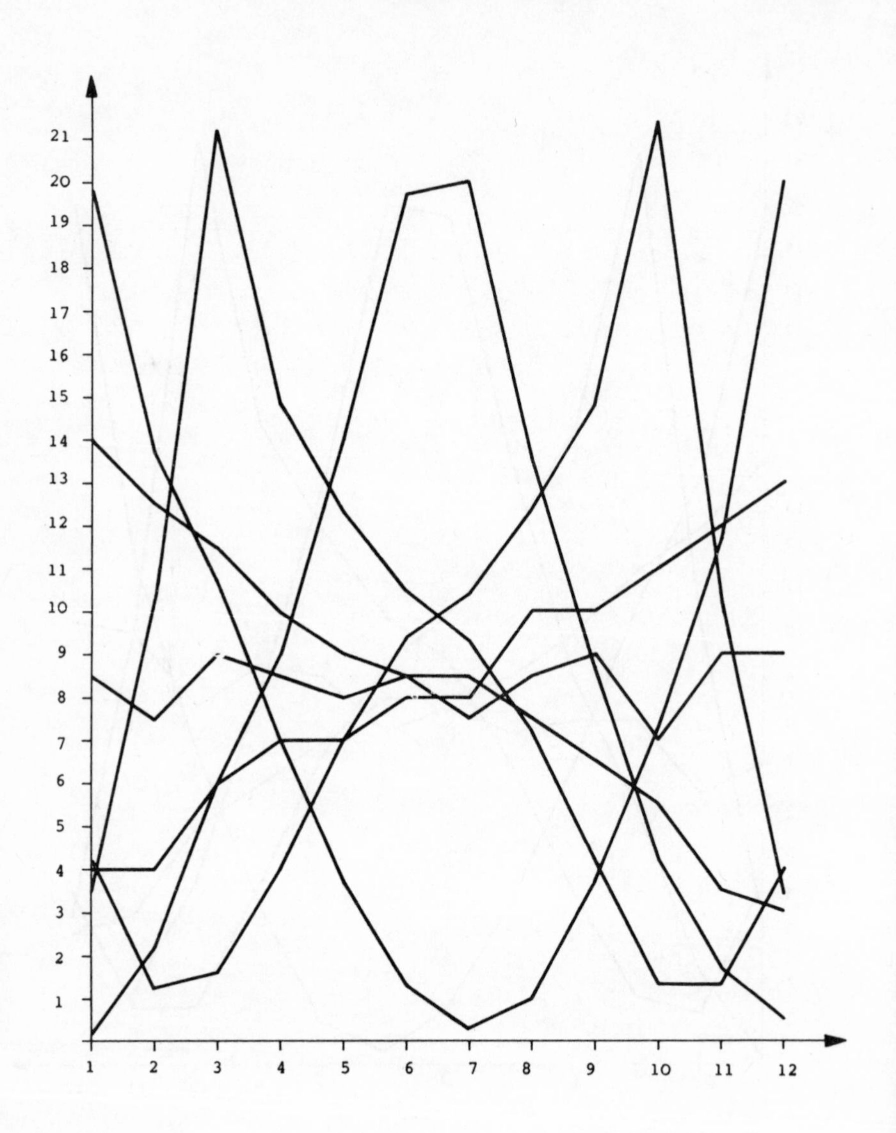

Fig. B 15

The following examples are calculated by using main programs which are not identical with H6 and are not given here. In the following, double precision was also used for the floating point numbers, which necessitates a simple modification of KMEANS, namely, introducing the variables concerned with a DOUBLE PRECISION statement, and changing built-in standard functions like FLOAT into DFLOAT. With large values of m, the use of double precision is not always necessary, but, because of the frequent centroid transformations (3.2.16) and (3.2.20), it contributes to greater accuracy.

Fig. T 5

800	MUENCHEN	330.93	1306024	3946.5
801	MUENCHEN-LAND OST	759.12	138706	182.7
802	MUENCHEN-LAND SUED	209.38	55749	266.3
803	MUENCHEN-LAND WEST	315.05	135167	429.0
804	MUENCHEN-LAND NORD	181.65	40366	222.2
805	FREISING	1085.39	131198	120.9
806	DACHAU	1019.17	121499	119.2
807	INGOLSTADT	576.91	141663	245.6
808	FUERSTENFELDBRUCK	365.02	54456	149.2
809	WASSERBURG	568.54	47490	83.5
810	GARMISCH-PARTENKIRCHEN	568.70	53483	94.0
811	MURNAU	459.37	27026	58.8
812	WEILHEIM	582.19	60539	104.0
813	STARNBERG	229.36	38607	168.3
815	HOLZKIRCHEN	322.42	23234	72.1
816	MIESBACH	347.73	30889	88.8
817	BAD TOELZ	671.30	36300	54.1
818	TEGERNSEE	271.42	21515	79.3
819	WOLFRATSHAUSEN	261.44	42482	162.5
820	ROSENHEIM	961.91	154739	160.9
821	PRIEN	441.50	39590	89.7
822	TRAUNSTEIN	1085.09	135459	124.8
823	BAD REICHENHALL	97.93	23578	240.8
824	BERCHTESGADEN	125.10	24836	198.5
825	DORFEN	505.88	38568	76.2
826	MUEHLDORF	1207.38	174628	144.6
830	LANDSHUT	1336.79	152268	113.9
831	LANDSHUT	1055.09	79532	75.4
833	EGGENFELDEN	456.25	33033	72.4
834	PFARRKIRCHEN	474.83	37083	78.1
835	PLATTLING	1692.89	156557	92.5
836	DEGGENDORF	24.49	20087	820.2
837	REGEN	858.06	74383	86.7
838	LANDAU (ISAR)	528.88	40814	77.2
839	PASSAU	1744.45	202356	116.0
840	REGENSBURG	679.88	188916	277.9
841	REGENSBURG	1123.26	109480	97.5
842	KELHEIM	744.65	67574	90.7
843	NEUMARKT	1114.27	91987	82.6
844	STRAUBING	973.13	106762	109.7
845	AMBERG	1199.54	136177	113.5
846	SCHWANDORF	593.69	61065	102.9
847	NABBURG	653.04	46071	70.5

848	WEIDEN	1053.85	142163	134.9
849	CHAM	1191.14	102431	86.0
850	NUERNBERG	984.24	685783	696.8
851	FUERTH	55.61	97563	1754.4
852	ERLANGEN	310.62	145511	468.5
853	NEUSTADT (AISCH)	889.59	68424	76.9
854	SCHWABACH	757.54	104125	137.5
855	FORCHHEIM	747.97	98102	131.2
856	LAUF	528.02	75965	143.9
857	PEGNITZ	432.77	41756	96.5
858	BAYREUTH	939.95	150127	159.7
859	MARKTREDWITZ	887.22	123161	138.8
860	BAMBERG	1726.77	225552	130.6
862	LICHTENFELS	538.57	89989	167.1
863	COBURG	453.58	110799	244.3
864	KRONACH	489.63	74806	152.8
865	KULMBACH	520.87	72111	138.4
866	MUENCHBERG	223.60	35439	158.5
867	HOF	900.46	190239	211.3
870	WUERZBURG	1282.67	270471	210.9
871	KITZINGEN	658.07	81060	123.2
872	SCHWEINFURT	1515.92	224789	148.3
873	BAD KISSINGEN	504.19	66731	132.4
874	BAD NEUSTADT (SAALE)	870.41	77400	88.9
875	ASCHAFFENBURG	710.54	238722	336.0
876	MILTENBERG	410.80	57350	139.6
877	LOHR	446.85	60078	134.4
878	GEMUENDEN	859.52	77689	90.4
880	ANSBACH	1672.24	174620	104.4
882	GUNZENHAUSEN	641.85	53229	82.9
883	TREUCHTLINGEN	1209.98	102429	84.7
885	DONAUWOERTH	1415.78	135007	95.4
886	NOERDLINGEN	539.99	51216	94.8
887	GUENZBURG	477.40	71767	150.3
888	DILLINGEN (DONAU)	561.72	62753	111.7
889	AICHACH	713.16	61575	86.3
890	AUGSBURG	1511.03	467367	309.3
891	LANDSBERG	577.75	60277	104.3
892	SCHONGAU	396.82	38152	96.1
893	SCHWABMUENCHEN	717.49	72692	101.3
894	MEMMINGEN	1026.50	123112	119.9
895	KAUFBEUREN	1048.81	120165	114.6
896	KEMPTEN	771.94	115852	150.1
897	IMMENSTADT	592.27	58453	98.7
898	OBERSTDORF	319.08	13348	41.8
899	LINDAU	323.48	69094	213.6

Fig. T 5

Fig. T5, from (Statistisches Bundesamt 1973) (German Federal Bureau for Statistics) gives the three-digit codes for postal zones in Bavaria, and their names. It also shows, as the variables $x_{.1}$ and $x_{.2}$, their surface areas in square kilometres, their populations, and the density of population $x_{.3} = x_{.2}/x_{.1}$. Figs. E6.4, E6.5, E6.6 and E6.7 show for the initial partition (3.2.10) using KMEANS, the $n = 7$

final partitions of the postal zones, their centroids and the sums of squared distances, using respectively the variables $x_{.1}$, $x_{.2}$, $(x_{.1}, x_{.2})$ and $x_{.3}$. The number of clusters $n = 7$ seemed to us outstandingly good in comparison with the results for $n = 3, \ldots, 15$. In E6.6 the two variables were scaled by means of TRANSF from Fig. U1 to give a mean of zero and standard deviation of one; in the other cases scalings are unnecessary, since only one variable appears. We leave the interested reader to interpret the cluster results using the two maps B16 and B17, clarified by the visual representation of the clusters in Figs. B18, B19, B20 and B21.

```
1 :  801  817  840  842  847  854  855  871  875  882  889  893
     896

2 :  805  822  826  830  831  841  843  845  848  849  870  883
     895

3 :  806  820  837  844  850  853  858  859  867  874  878  894

4 :  802  804  813  823  824  836  851  866

5 :  800  803  808  815  816  818  819  852  876  892  898  899

6 :  807  809  810  811  812  821  825  833  834  838  846  856
     857  862  863  864  865  873  877  886  887  888  891  897

7 :  835  839  860  872  880  885  890

   709.735      1153.328       930.847       143.390       331.234
   517.515      1611.297

    0.242D 05        0.102D 06        0.412D 05        0.439D 05
    0.215D 05        0.648D 05        0.988D 05        0.397D 06
```

Fig. E 6.4

```
1 :  804  809  811  813  815  816  817  818  819  821  823  824
     825  833  834  836  838  847  857  866  892  898

2 :  801  803  805  807  820  822  826  830  835  845  848  852
     858  880  885

3 :  850  890

4 :  839  840  860  867  870  872  875

5 :  802  808  810  812  831  837  842  846  853  856  864  865
     871  873  874  876  877  878  882  886  887  888  889  891
     893  897  899

6 :  806  841  843  844  849  851  854  855  859  862  863  883
     894  895  896

7 :  800


 33648.364  146932.667  576575.000  220149.286   65905.593
107830.400 1306024.000


   0.180D 10       0.262D 10       0.239D 11       0.511D 10
   0.212D 10       0.173D 10       0.529D-02       0.372D 11
```

Fig. E 6.5

```
1 :  801  807  837  840  842  853  854  855  859  871  874  875
     878  889  893  896

2 :  805  806  820  822  826  831  841  843  844  845  848  849
     858  867  883  894  895

3 :  850  890

4 :  800

5 :  802  803  804  808  813  815  816  818  819  823  824  836
     851  852  866  898  899

6 :  809  810  811  812  817  821  825  833  834  838  846  847
     856  857  862  863  864  865  873  876  877  882  886  887
     888  891  892  897

7 :  830  835  839  860  870  872  880  885


     0.011        0.092        0.139        -0.103        -0.128
    -0.053        0.218

    -0.005        0.008        0.311         0.805        -0.044
    -0.041        0.052


    0.251D-01       0.164D-01       0.206D-01       0.297D-14
    0.238D-01       0.143D-01       0.235D-01       0.124D 00
```

Fig. E 6.6

```
1 :  809  810  811  812  815  816  817  818  821  825  830  831
     833  834  835  837  838  839  841  842  843  844  845  846
     847  849  853  857  874  878  880  882  883  885  886  888
     889  891  892  893  895  897  898

2 :  802  804  807  823  824  840  863  867  870  875  890  899

3 :  836  850

4 :  851

5 :  800

6 :  803  852

7 :  801  805  806  808  813  819  820  822  826  848  854  855
     856  858  859  860  862  864  865  866  871  872  873  876
     877  887  894  896


    89.208      248.039      758.488      1754.415      3946.526
   448.743      143.739


    0.112D 05        0.198D 05        0.762D 04        0.795D-07
    0.188D-06        0.777D 03        0.721D 04        0.467D 05
```

Fig. E 6.7

874
863
864
873
866
867
878
862
865
875
877
872
860
858
859
870
86
87
876
871
855
857
848
853
852
856
850
85
845
847
880
846
843
849
854
882
841
84
837
883
840
844
886
842
836
807
835
885
830
839
888
838
83
889
806
831
887
890
89
805
833
834
825
826
808
8
803
893
801
809
894
891
802
813
822
819
815
812
895
820
821
896
892
816
899
811
817
818
823
897
824
810
898

Fig. B 16

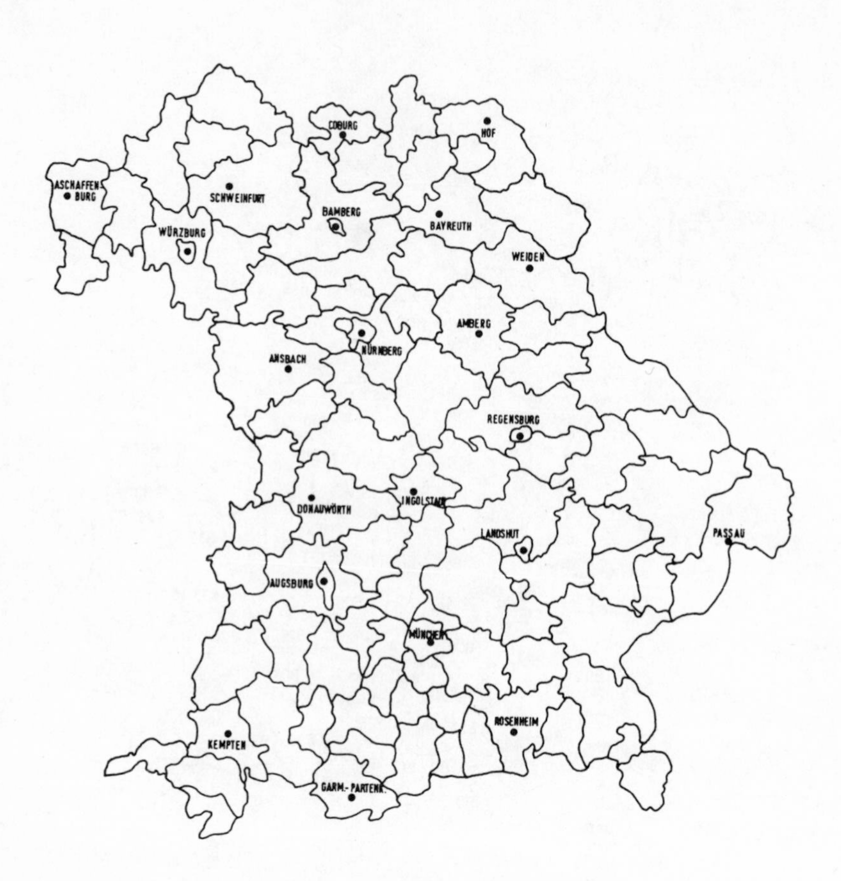
COBURG
HOF
ASCHAFFEN-
BURG
SCHWEINFURT
BAMBERG
BAYREUTH
WÜRZBURG
WEIDEN
AMBERG
NÜRNBERG
ANSBACH
REGENSBURG
INGOLSTADT
DONAUWÖRTH
LANDSHUT
PASSAU
AUGSBURG
MÜNCHEN
ROSENHEIM
KEMPTEN
GARM.- PARTENK.

Fig. B 17

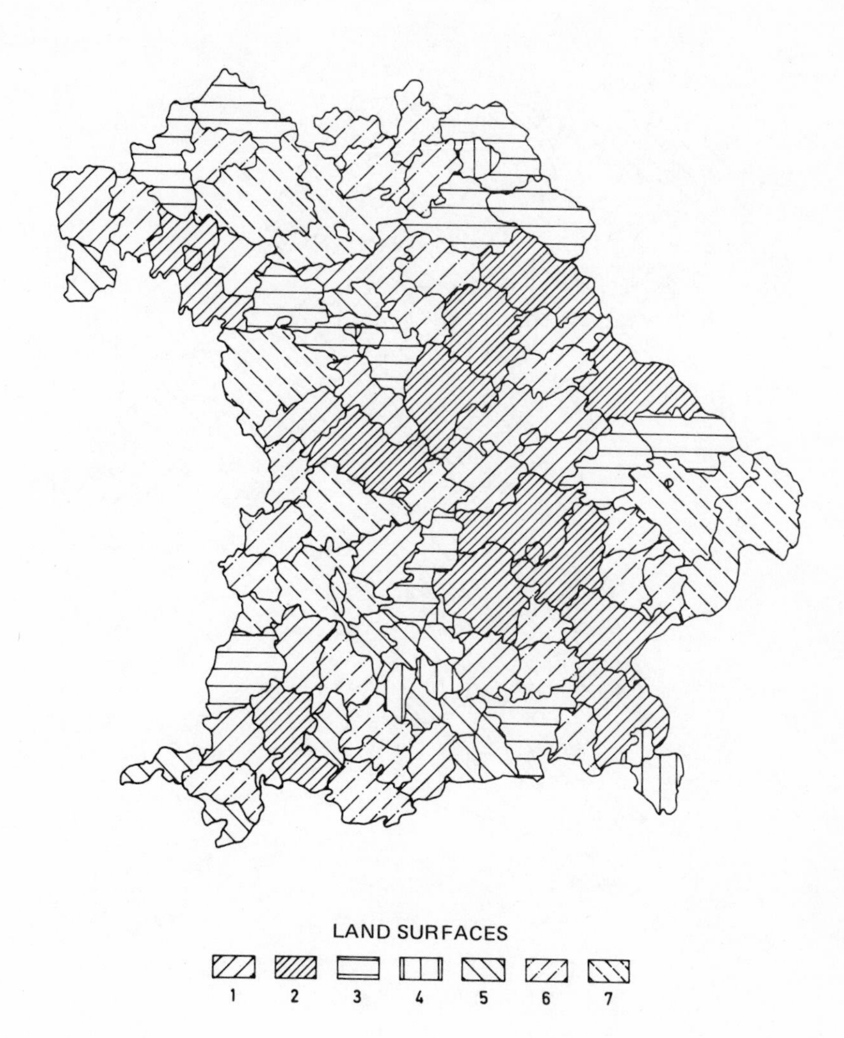

Fig. B 18

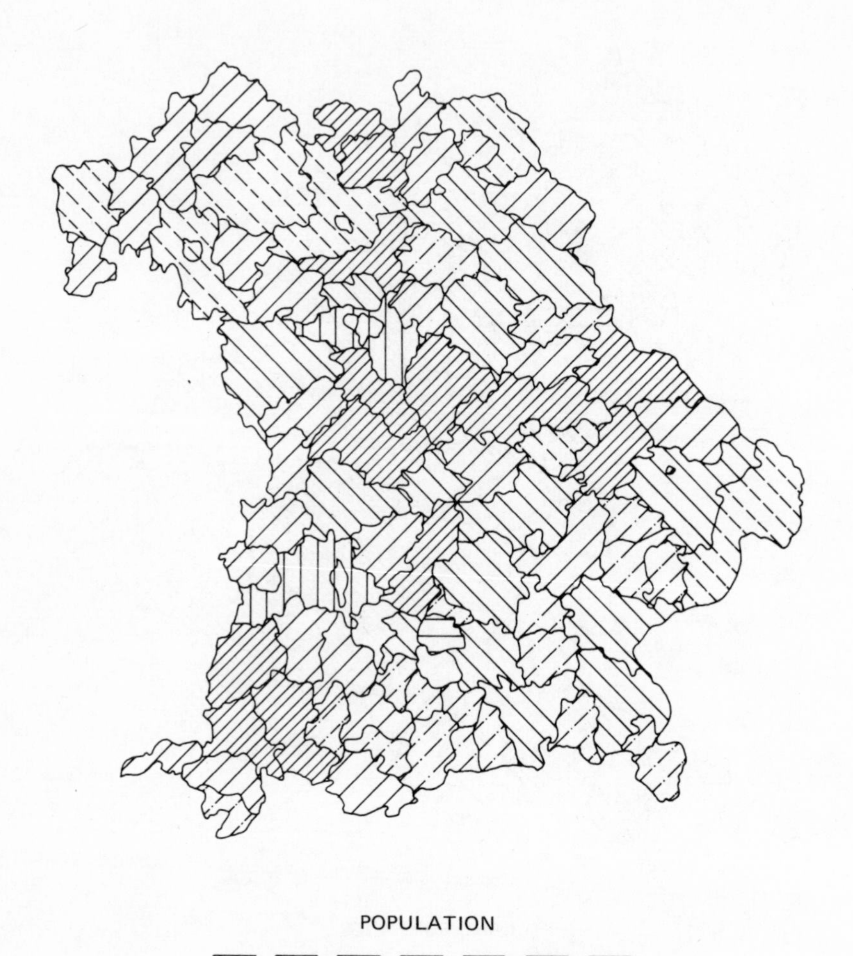

Fig. B 19

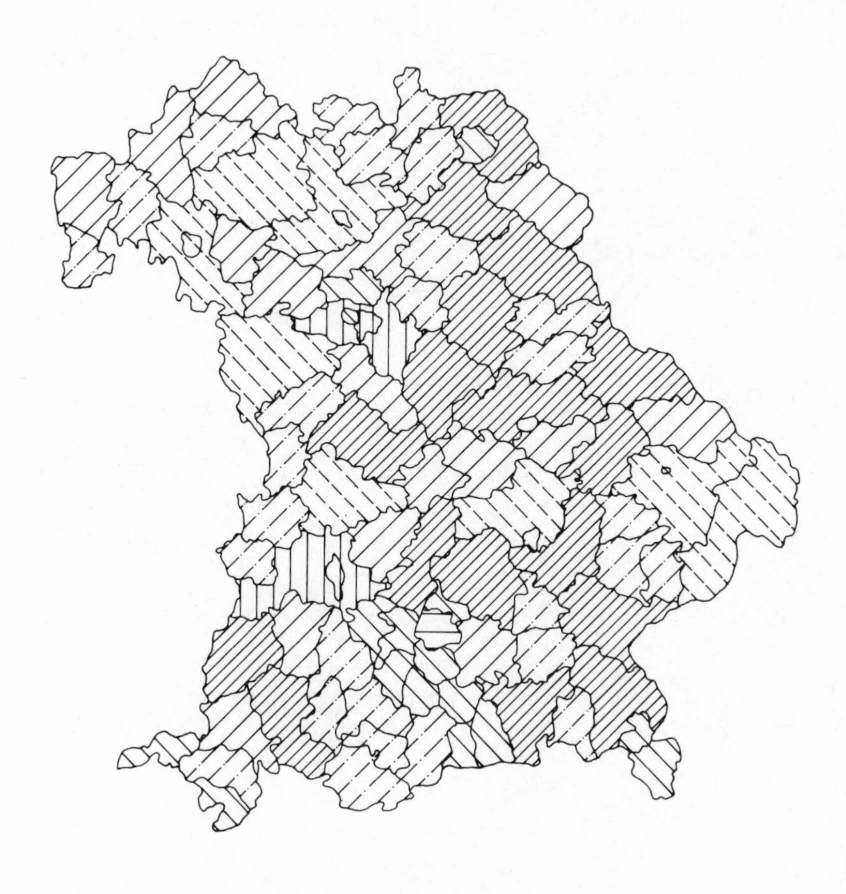

LAND SURFACES AND POPULATION

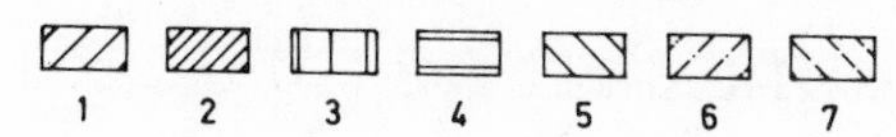

Fig. B 20

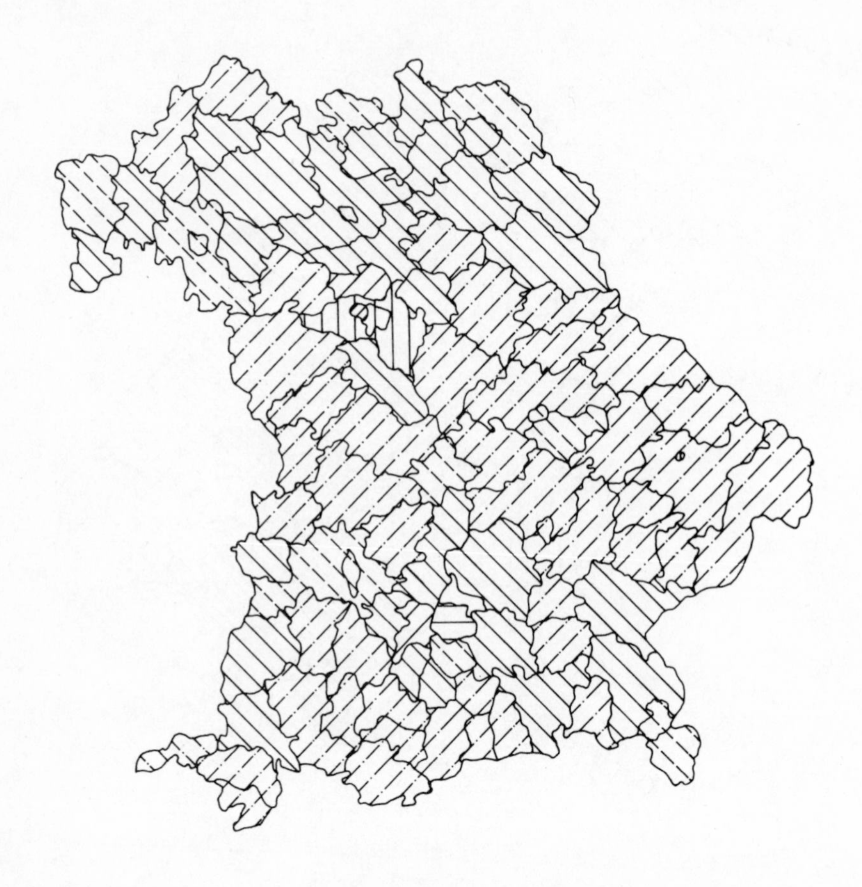

DENSITY OF POPULATION

Fig. B 21

For practical purposes there will be other meaningful variables such as, say, size of household, average income or educational opportunities, according to the purpose to which one's efforts are directed, within postal zones, other area units or even cities. Demographic variables, which make it possible to employ sales representatives for agreed clusters of postal districts or municipalities, according to capabilities, mental attitudes and sales techniques, are all worthy of consideration (see Howard 1966). The possibilities opened up are legion.

Fig. T 6

800	MUENCHEN	56750	57218	300201	242375
801	MUENCHEN-LAND OST	7684	5790	20279	23491
802	MUENCHEN-LAND SUED	3780	1977	11058	7398
803	MUENCHEN-LAND WEST	7226	5623	25571	20380
804	MUENCHEN-LAND NORD	2226	1305	9347	12432
805	FREISING	8187	5140	14632	24377
806	DACHAU	8165	2763	11638	24489
807	INGOLSTADT	5810	5212	15019	30532
808	FUERSTENFELDBRUCK	3017	3045	6733	9412
809	WASSERBURG	3942	1153	3119	8220
810	GARMISCH-PARTENKIRCHEN	3972	2766	7301	8909
811	MURNAU	1991	932	2752	4912
812	WEILHEIM	3499	1687	5608	11470
813	STARNBERG	2469	1751	6492	5112
815	HOLZKIRCHEN	2070	598	2674	4820
816	MIESBACH	2291	836	2791	5688
817	BAD TOELZ	2587	927	3960	6469
818	TEGERNSEE	2075	454	3124	3435
819	WOLFRATSHAUSEN	1663	504	3599	6885
820	ROSENHEIM	9438	5778	16163	28787
821	PRIEN	2920	1143	3483	7025
822	TRAUNSTEIN	8747	4844	12417	25132
823	BAD REICHENHALL	1614	1541	2792	3683
824	BERCHTESGADEN	1684	1136	2851	4810
825	DORFEN	3757	652	2161	7287
826	MUEHLDORF	10126	4043	14208	32746
830	LANDSHUT	10171	5760	13948	25430
831	LANDSHUT	7259	1484	4638	14979
833	EGGENFELDEN	3280	578	1607	5057
834	PFARRKIRCHEN	3867	835	1745	5757
835	PLATTLING	11764	3506	7203	27350
836	DEGGENDORF	978	1472	2726	3321
837	REGEN	4322	2064	3432	14601
838	LANDAU (ISAR)	4148	830	1948	6706
839	PASSAU	12628	6745	13733	36240
840	REGENSBURG	7973	10559	24144	31991
841	REGENSBURG	5884	3122	6775	23835
842	KELHEIM	4417	1485	3922	13172
843	NEUMARKT	6128	2192	5550	18791
844	STRAUBING	8065	3942	7945	17942
845	AMBERG	6331	4737	10942	26253
846	SCHWANDORF	2804	2720	3448	11911
847	NABBURG	3078	1976	2195	8889
848	WEIDEN	7061	5029	10980	28747
849	CHAM	7740	3676	4994	17862
850	NUERNBERG	26467	23372	114459	142258

851	FUERTH	3751	2743	17017	25602
852	ERLANGEN	4815	4252	25409	26261
853	NEUSTADT (AISCH)	5503	1536	4260	13437
854	SCHWABACH	6059	3405	9951	21382
855	FORCHHEIM	5843	2164	7288	21638
856	LAUF	4315	1942	7505	17506
857	PEGNITZ	2435	878	2473	9316
858	BAYREUTH	8250	6385	15099	28740
859	MARKTREDWITZ	6468	3331	9463	29032
860	BAMBERG	12315	7007	18766	44127
862	LICHTENFELS	5139	2330	7652	23557
863	COBURG	5035	3803	11789	26103
864	KRONACH	3664	1928	5152	19647
865	KULMBACH	3996	1812	6614	16570
866	MUENCHBERG	2263	751	3172	9080
867	HOF	8924	5485	18693	48969
870	WUERZBURG	12764	12361	32747	39394
871	KITZINGEN	6036	1791	6405	13785
872	SCHWEINFURT	11262	5030	20789	43324
873	BAD KISSINGEN	3612	2292	5552	14036
874	BAD NEUSTADT (SAALE)	4629	2462	5255	15462
875	ASCHAFFENBURG	8933	5669	24468	54208
876	MILTENBERG	2604	1209	4653	13266
877	LOHR	2827	1524	4974	12239
878	GEMUENDEN	3962	3759	5378	14633
880	ANSBACH	12833	6077	13434	32520
882	GUNZENHAUSEN	4509	1225	2993	9563
883	TREUCHTLINGEN	6720	3731	6679	20093
885	DONAUWOERTH	10014	5104	9125	25162
886	NOERDLINGEN	4245	1204	3241	8592
887	GUENZBURG	4260	2491	5648	15665
888	DILLINGEN (DONAU)	4402	1617	4641	13225
889	AICHACH	4615	1285	3667	12983
890	AUGSBURG	19951	14561	63228	101176
891	LANDSBERG	4482	3696	6893	12158
892	SCHONGAU	2627	1194	2803	8037
893	SCHWABMUENCHEN	6044	2520	5569	14411
894	MEMMINGEN	9575	3439	11234	22490
895	KAUFBEUREN	8508	4479	11723	22118
896	KEMPTEN	7661	3951	13258	20624
897	IMMENSTADT	4163	2505	5871	12812
898	OBERSTDORF	1375	404	1653	2192
899	LINDAU	4409	2452	7370	13256

Fig. T 6

We shall give another example for the above Bavarian postal zones, namely classification according to specific aspects of the population (see Bayerisches Statistisches Landesamt (Bavarian Provincial Statistics Bureau) 1970, Lades 1974). This time we take as variables the numbers of self-employed people, of civil servents, clerks and manual workers; these are to be found in this order in Fig. T6. The values of the variables are taken from the 1970 population census and again were standardised by means of TRANSF, before the application of KMEANS, to have means of zero and standard deviations of one. The results are given in Figs. E6.8 to E6.14, this time for $n = 3, \ldots, 9$ with initial partitions given by (3.2.10). Here again we leave to the reader the interpretation of the clusters and, on this occasion, also the selection of a satisfactory value for the number of clusters. It is clear that the centroids make the interpretation easier, particularly if one calculates them on the basis of the original scaling.

```
1 :  839  840  850  860  867  870  872  875  880  890

2 :  800

3 :  801  802  803  804  805  806  807  808  809  810  811  812
     813  815  816  817  818  819  820  821  822  823  824  825
     826  830  831  833  834  835  836  837  838  841  842  843
     844  845  846  847  848  849  851  852  853  854  855  856
     857  858  859  862  863  864  865  866  871  873  874  876
     877  878  882  883  885  886  887  888  889  891  892  893
     894  895  896  897  898  899

    0.1101        0.8002        -0.0244

    0.0927        0.8626        -0.0229

    0.0655        0.9034        -0.0200

    0.1219        0.7714        -0.0255

   0.3856D 00     0.7500D-13     0.2509D 00     0.6365D 00
```

Fig. E 6.8

```
1 :  802  804  808  809  810  811  812  813  815  816  817  818
     819  821  823  824  825  831  833  834  836  837  838  841
     842  843  846  847  851  853  855  856  857  862  864  865
     866  871  873  874  876  877  878  882  886  887  888  889
     891  892  893  897  898  899

2 :  850  890

3 :  801  803  805  806  807  820  822  826  830  835  839  840
     844  845  848  849  852  854  858  859  860  863  867  870
     872  875  880  883  885  894  895  896

4 :  800

   -0.0452      0.2661      0.0346      0.8002

   -0.0369      0.2430      0.0202      0.8626

   -0.0278      0.2370      0.0039      0.9034

   -0.0402      0.3477      0.0220      0.7714


   0.5874D-01      0.3901D-01      0.1130D 00      0.3238D-13
   0.2107D 00
```

Fig. E 6.9

```
1 :  800

2 :  801  803  805  806  807  822  831  841  843  844  845  848
     849  851  852  854  855  859  862  863  883  893  894  895
     896

3 :  820  826  830  835  839  840  858  860  867  870  872  875
     880  885

4 :  802  804  808  809  810  811  812  813  815  816  817  818
     819  821  823  824  825  833  834  836  837  838  842  846
     847  853  856  857  864  865  866  871  873  874  876  877
     878  882  886  887  888  889  891  892  897  898  899

5 :  850  890

     0.8002       0.0048       0.0642      -0.0501       0.2661

     0.8626      -0.0035       0.0394      -0.0386       0.2430

     0.9034      -0.0067       0.0115      -0.0292       0.2370

     0.7714       0.0004       0.0455      -0.0450       0.3477

   0.2726D-13     0.3317D-01     0.4795D-01     0.3539D-01
   0.3901D-01     0.1555D 00
```

Fig. E 6.10

```
1 :  802  804  808  809  811  812  813  815  816  817  818  819
     821  823  824  825  833  834  836  838  846  847  857  866
     876  877  882  886  892  898

2 :  800

3 :  801  803  805  806  807  820  822  826  830  835  844  845
     848  849  852  858  859  883  885  894  895  896

4 :  850  890

5 :  810  831  837  841  842  843  851  853  854  855  856  862
     863  864  865  871  873  874  878  887  888  889  891  893
     897  899

6 :  839  840  860  867  870  872  875  880


   -0.0595      0.8002      0.0258      0.2661      -0.0263
    0.0710

   -0.0445      0.8626      0.0098      0.2430      -0.0258
    0.0551

   -0.0313      0.9034     -0.0018      0.2370      -0.0223
    0.0227

   -0.0536      0.7714      0.0078      0.3477      -0.0213
    0.0655


   0.1358D-01     0.2183D-13     0.2958D-01     0.3901D-01
   0.1759D-01     0.2826D-01     0.1280D 00
```

Fig. E 6.11

```
1 :   804  809  811  813  815  816  817  818  819  821  823  824
      825  833  834  836  838  847  857  866  876  877  886  892
      898

2 :   820  822  826  830  835  858  880  885  894

3 :   850  890

4 :   839  840  860  867  870  872  875

5 :   802  808  810  812  831  837  842  843  846  851  853  855
      856  862  864  865  871  873  874  878  882  887  888  889
      891  893  897  899

6 :   801  803  805  806  807  841  844  845  848  849  852  854
      859  863  883  895  896

7 :   800


    -0.0620      0.0575      0.2661      0.0668      -0.0314
     0.0066      0.8002

    -0.0475      0.0167      0.2430      0.0581      -0.0283
     0.0044      0.8626

    -0.0327     -0.0035      0.2370      0.0260      -0.0237
    -0.0028      0.9034

    -0.0554      0.0172      0.3477      0.0699      -0.0282
     0.0037      0.7714


    0.8977D-02      0.8558D-02      0.3901D-01      0.2502D-01
    0.1792D-01      0.1712D-01      0.1256D-13      0.1166D 00
```

Fig. E 6.12

```
1 :  801  803  807  845  848  851  852  859  863

2 :  802  808  810  812  837  842  846  853  856  864  865  873
     874  878  882  886  887  888  889  891  897  899

3 :  806  831  841  843  844  849  854  855  862  871  883  893
     896

4 :  805  820  822  826  830  835  858  885  894  895

5 :  804  809  811  813  815  816  817  818  819  821  823  824
     825  833  834  836  838  847  857  866  876  877  892  898

6 :  800

7 :  839  840  860  867  870  872  875  880

8 :  850  890


   -0.0075      -0.0378      0.0029       0.0475      -0.0631
    0.8002       0.0710      0.2661

    0.0087      -0.0289     -0.0180       0.0143      -0.0476
    0.8626       0.0551      0.2430

    0.0082      -0.0257     -0.0192      -0.0034      -0.0327
    0.9034       0.0227      0.2370

    0.0125      -0.0340     -0.0112       0.0124      -0.0556
    0.7714       0.0655      0.3477


   0.9665D-02      0.7794D-02      0.7927D-02      0.6977D-02
   0.8220D-02      0.1521D-13      0.2826D-01      0.3901D-01
   0.1079D 00
```

Fig. E 6.13

```
1 :  807  845  848  851  852  859  863

2 :  839  860  867  870  872  875  880

3 :  805  820  822  826  830  835  885  894  895

4 :  806  831  841  843  844  849  854  855  862  871  883  893
     896

5 :  801  803  840  858

6 :  802  808  810  812  837  842  846  853  856  864  865  873
     874  878  882  886  887  888  889  891  897  899

7 :  804  809  811  813  815  816  817  818  819  821  823  824
     825  833  834  836  838  847  857  866  876  877  892  898

8 :  800

9 :  850  890


    -0.0141      0.0778      0.0497      0.0029      0.0205
    -0.0378     -0.0631      0.8002      0.2661

     0.0032      0.0477      0.0116     -0.0180      0.0506
    -0.0289     -0.0476      0.8626      0.2430

     0.0022      0.0212     -0.0043     -0.0192      0.0240
    -0.0257     -0.0327      0.9034      0.2370

     0.0169      0.0702      0.0115     -0.0112      0.0121
    -0.0340     -0.0556      0.7714      0.3477


    0.5390D-02     0.2127D-01     0.5707D-02     0.7927D-02
    0.6096D-02     0.7794D-02     0.8220D-02     0.1184D-13
    0.3901D-01     0.1014D 00
```

Fig. E 6.14

As our final example we take the 64 main two-digit codes for the postal zones of the Federal German Republic, and interpret the three variables (which have already been standardised to a mean of zero and standard deviation of one) in columns four to six in E6.1 respectively, as the turnover of three car firms, or as the turnover of three retail outlets of a diversified company. Table T7 contains the values $Z_1 = d$ for various numbers of clusters using the initial partitions (3.2.10):

Table T7

n	2	3	4	5	6	7	8	9	10
d	110.14	75.31	51.48	44.63	37.67	25.08	20.78	17.48	16.45

On the basis of this table and also of the number $m = 64$, it now seems to us sensible to select $n = 7$. The corresponding vector p_i $(i = 1, \ldots, 64)$ is given in the second column of E6.15 and the corresponding postal codes are shown in the third. Fig. T8 shows the cluster centroids of the three components and the sums of squared distances for $n = 7$.

Fig. E 6.15

			1	2	3
1	3	10	-0.15	0.23	2.76
2	3	20	0.46	0.11	0.79
3	7	21	-0.43	-0.57	-0.57
4	7	22	-0.43	-0.57	-0.47
5	5	23	0.58	0.87	-0.28
6	2	24	-1.31	-0.96	-0.52
7	5	28	0.64	0.79	-0.14
8	7	29	0.04	-0.84	-0.30
9	3	30	0.36	0.20	0.53
10	7	31	0.39	-0.12	-0.32
11	2	32	-0.75	-0.77	-0.39
12	5	33	0.63	0.20	-0.26
13	7	34	-0.34	-0.30	-0.49
14	6	35	1.50	1.60	-0.16
15	1	40	-0.54	-0.72	2.24
16	1	41	-0.77	-1.11	0.80
17	2	42	-1.33	-1.50	-0.44
18	1	43	-0.92	-0.89	0.89
19	7	44	-0.29	-0.63	-0.28
20	2	45	-1.32	-1.37	-0.48
21	1	46	-0.66	-0.82	1.11
22	2	47	-2.03	-1.85	0.06
23	2	48	-0.67	-0.63	-0.38
24	2	49	-0.78	-0.76	-0.57
25	7	50	0.68	-0.75	-0.34
26	7	51	0.01	0.10	-0.32

27	7	52	0.05	-0.34	-0.56
28	2	53	-1.05	-1.03	-0.52
29	4	54	0.34	1.24	5.42
30	7	55	-0.67	-0.29	-0.48
31	7	56	-0.37	-0.75	-0.53
32	2	57	-1.47	-1.22	-0.58
33	1	58	-1.37	-0.92	1.01
34	7	59	0.18	-0.06	-0.34
35	7	60	-0.40	-0.35	-0.48
36	1	61	-1.05	-0.81	1.33
37	7	62	0.18	0.36	-0.69
38	5	63	0.72	0.48	-0.37
39	5	64	0.90	0.63	-0.47
40	7	65	-0.33	-0.04	0.49
41	5	66	1.21	1.15	-0.19
42	6	67	1.57	2.18	-0.31
43	2	68	-1.22	-1.29	-0.27
44	7	69	0.37	0.02	-0.69
45	6	70	1.98	1.43	-0.26
46	6	71	1.43	1.57	-0.69
47	7	72	-0.11	-0.31	-0.59
48	2	73	-0.89	-0.75	-0.69
49	7	74	0.02	0.04	-0.56
50	5	75	0.57	0.68	-0.28
51	2	76	-1.41	-1.05	-0.53
52	2	77	-0.67	-0.97	-0.63
53	7	78	0.11	0.19	-0.54
54	5	79	0.60	0.97	-0.39
55	6	80	3.41	1.47	-0.09
56	2	81	-0.59	-0.79	-0.69
57	5	82	1.22	1.41	-0.67
58	5	83	0.58	0.66	-0.50
59	3	84	0.72	1.59	0.95
60	3	85	-0.19	0.17	1.15
61	5	86	0.57	1.75	-0.29
62	6	87	2.48	2.83	-0.30
63	7	88	-0.16	0.37	-0.00
64	3	89	0.16	0.79	1.36

Fig. E 6.15

Table T8

Cluster	Centroids	e
1	(−.88, −.88, 1.23)	1.93
2	(−1.11, −1.07, −.47)	4.19
3	(.23, .52, 1.26)	5.42
4	(.34, 1.24, 5.42)	0
5	(.75, .87, −.35)	2.76
6	(2.06, 1.85, −.30)	4.70
7	(−.08, −.24, −.40)	6.08

It is clear that on the basis of the cluster results each car firm could be shown how it stands quantitatively relative to the others in the various types of zone. Similarly the company could assess the relationship between its three outlets and, in combination with demographic data or attributes, identify potential openings in the market. To round off, the number of passes necessary, and the number of points moved at each stage, are shown in Table T9, for numbers of clusters $n = 2, \ldots, 10$.

Table T9

n\Pass	1	2	3	4	5	6
2	33	2				
3	39	4	2			
4	42	13	5	2	2	
5	49	13	1	4		
6	53	20	14	2	5	3
7	46	16	11	7	6	
8	49	21	4	4	5	1
9	56	19	7	8		
10	55	15	11	10	2	

We shall now discuss several variations of KMEANS in cooperation with other heuristic clustering techniques which are well worth thinking about and which are mentioned, in part, in the literature available.

(Sparks 1973) gives a Fortran algorithm nearly identical to KMEANS. This presumes an initial set of centroids instead of an initial partitioning. Each point is assigned to the nearest centroid and the exchange process triggered. A minimal number of points per cluster, which in KMEANS is 1, can be prescribed.

A variation of the exchange process suggested in (Friedman and Rubin 1967) is that of dealing with the points in clusters instead of running through them in turn from $i = 1, \ldots, m$. This may not alter the program other than to apply more time-consuming logic to it.

There are several suggestions in (Ball 1965, Ball and Hall 1967, MacQueen 1967) which advise against rigid adherence to the assumed number of clusters, but to split into two clusters, after each pass, any cluster for which the sum of squares e_j exceeds some threshold value, or to split up into two clusters only that cluster which has the greatest value e_j. In the same way it is suggested that pairs of clusters of which the separation is, respectively, $\|\bar{x}_j - \bar{x}_k\|$ or $\frac{m_j m_k}{m_j + m_k} \|\bar{x}_j - \bar{x}_k\|$ and the separation of which does not reach the suggested threshold – or the two clusters with the very smallest separation – should be united. In this way poor local optima, which might otherwise be reached, ought to be avoided. Maximum and minimum numbers of elements could also be substituted for the two threshold values. In this case, n does or does not alter.

Since no clusters of only one element normally appear in the final partitioning produced by KMEANS when satisfactory values of m/n – roughly $m/n \geqslant 5$ – are used, we have not implemented any of the methods described for dividing and merging. What these do, in fact, is replace choice of the number of clusters by choice of threshold values for e_j or for the number of elements. Again, we recommend for KMEANS calculating with various values of the number of clusters and with several random initial partitions (see Späth 1977b).

Another variation worth considering would be that using the formulae (3.2.15) and (3.2.19) where a group of objects is transferred experimentally from one cluster to another. This raises such questions as how to choose the objects.

Incidentally, one can try to simplify the KMEANS principle so that a point is not always transferred to that cluster in which it reduces the value of the objective function by the greatest amount, but to the first available cluster in which the objective function would be lowered (see Fisher and Van Ness 1971). Empirically this less cumbersome method seems to behave less well than the stepwise optimal exchange method (see Späth 1977c).

A great advantage of the KMEANS principle is that, with a pre-defined data matrix (1.1) whose rows are processed in sequence, KMEANS can be reorganised so that these rows can be stored sequentially on an external medium and also processed sequentially at each pass. As a result there need be no practical limitation on m and l because of storage requirements (see Anderberg 1973).

We now apply the objective function and the KMEANS principle to the case in which a symmetrical distance matrix, determined in some way, is provided instead of a metric data matrix.

By means of (3.1.8) the objective function Z_1 can be converted into

$$Z_1(C_1,\ldots,C_n) = \sum_{j=1}^{n} \frac{1}{m_j} \sum_{i\in C_j} \sum_{\substack{k\in C_j \\ k>i}} \|x_i - x_k\|^2 \longrightarrow \text{Minimum} \qquad (3.2.26)$$

Here the centroids do not appear. The distances $d_2^2(x_i, x_k) = \|x_i - x_k\|^2$ $(i, k = 1, \ldots, m)$ could be calculated once at the start and then the exchange algorithm can operate exactly like KMEANS in a modified form. In (Jensen 1969) dynamic programming (see Duran and Odell 1974) is used to minimise (3.2.26). This is certainly more effective than full enumeration, but is less effective by far than the KMEANS principle, which of course, does not guarantee a global optimum. If, by analogy with (3.2.12), one sets

$$e_j = \frac{1}{m_j} \sum_{i \in C_j} \sum_{\substack{k \in C_j \\ k > i}} \|x_i - x_k\|^2, \tag{3.2.27}$$

then for $C_p = C_j - \{k\}$

$$e_p = \frac{m_j e_j - \sum_{i \in C_j} \|x_k - x_i\|^2}{m_j - 1} \tag{3.2.28}$$

and for $C_p = C_j \cup \{k\}$

$$e_p = \frac{m_j e_j + \sum_{i \in C_j} \|x_k - x_i\|^2}{m_j + 1}. \tag{3.2.29}$$

both follow. These formulae correspond to (3.2.17) and (3.3.21). As in (3.2.22), one now tries to determine in several passes whether there is at least one $j \neq r$ for which

$$e_r - e_p > e_q - e_j \quad (C_p = C_r - \{i\},\ C_q = C_j \cup \{i\}). \tag{3.2.30}$$

is true for a point x_i with $i \in C_r$. If no such j is found, the next point is taken in its turn. Otherwise x_i is assigned to that cluster, with the serial number u, for which the right-hand side is smallest. The objective function Z_1 is therefore reduced at this step by the amount

$$e_r - e_p + e_u - e_q \quad (q = C_u \cup \{i\}). \tag{3.2.31}$$

The transformation of Z_1 has been carried out for Euclidean distances. However, arbitrary and non- metric distance functions can be used in (3.2.26), writing accordingly

$$Z_1(C_1, \ldots, C_n) = \sum_{j=1}^{n} \frac{1}{m_j} \sum_{i \in C_j} \sum_{\substack{k \in C_j \\ k > i}} d^2(x_i, x_k) \tag{3.2.32}$$

In the corresponding interpretation of e_j the algorithm operates analogously, even when the square on the right-hand side is omitted and, under the assumption that $d(x_i, x_k) \geqslant 0$, one writes

$$Z_1(C_1, \ldots, C_n) = \sum_{j=1}^{n} \frac{1}{m_j} \sum_{i \in C_j} \sum_{\substack{k \in C_j \\ k > i}} d(x_i, x_k) \tag{3.2.33}$$

In addition the weight $1/m_j$ can be omitted, giving

$$Z_1(C_1, \ldots, C_n) = \sum_{j=1}^{n} \sum_{i \in C_j} \sum_{\substack{k \in C_j \\ k > i}} d(x_i, x_k) \tag{3.2.34}$$

Modifications of (3.2.28) and (3.2.29) are necessary here. The KMEANS principle can likewise be modified to apply to given similarity coefficients $s(x_i, x_k)$ (see Fortier and Solomon 1966), to maximise

$$Z_1(C_1, \ldots, C_n) = \sum_{j=1}^{n} \sum_{i \in C_j} \sum_{\substack{k \in C_j \\ k > i}} s(x_i, x_k) \tag{3.2.35}$$

For this, (3.2.30) has to be modified correspondingly.

Fig. U 8

```
      SUBROUTINE CLUDIA (M,T,P,N,E,D,IDR)
C
C     CLUSTERING OF SYMMETRICAL MATRICES T(M,M) ACCORDING TO
C     THE KMEANS PRINCIPLE.
C
C     LET D = SUM (E(J)) WITH E(J) = SUM(SUM(T(I,K)))/Q(J)
C                                         I   K>I
C     WHERE P(I) = P(K) = J.
C
C     FOR AN INITIAL ALLOCATION P(I), WITH 1 <= P(I) <= N AND
C     NON-EMPTY CLUSTERS, D IS REDUCED AS FAR AS POSSIBLE
C     BY REPEATED EXCHANGE OF CLUSTER MEMBERS, AND
C     P(I) (I = 1,...,M) MODIFIED.   IF IDR IS NON-ZERO,
C     INTERMEDIATE RESULTS ARE PRINTED.
C
C     DIMENSION T(M,M),  P(M), S(M-N),E(N), Q(N), C(N)
      DIMENSION T(60,60),P(60),S(60)  ,E(10),Q(10),C(10)
      INTEGER   P,Q,R,R1,S,H,U,V
      KO=6
      D=0.
      DO 5 J=1,N
            R=0
            E(J)=0.
            DO 1 I=1,M
                  IF(P(I).NE.J) GOTO 1
                  R=R+1
                  S(R)=I
1           CONTINUE
            R1=R-1
            IF(R1.EQ.0) GOTO 4
            F=0.
            DO 3 K=1,R1
                  K1=K+1
                  U=S(K)
                  DO 2 L=K1,R
                        V=S(L)
                        F=F+T(U,V)
```

```
2          CONTINUE
3       CONTINUE
        F=F/FLOAT(R)
        E(J)=F
        D=D+F
4       Q(J)=R
5  CONTINUE
   I=0
   IT=0
6  I=I+1
   IF(I.GT.M) I=I-M
   IF(IT.EQ.M) RETURN
   R=P(I)
   U=Q(R)
   IF(U.LE.1) GOTO 6
   F=FLOAT(U)
   DO 7 J=1,N
        C(J)=0.
7  CONTINUE
   DO 8 H=1,M
        IF(H.EQ.I) GOTO 8
        V=P(H)
        C(V)=C(V)+T(I,H)
8  CONTINUE
   A=(F*E(R)-C(R))/(F-1.)
   AA=E(R)-A
   BB=1.E30
   DO 9 J=1,N
        IF(J.EQ.R) GOTO 9
        F=FLOAT(Q(J))
        Y=(F*E(J)+C(J))/(F+1.)
        YY=Y-E(J)
        IF(YY.GE.BB) GOTO 9
        BB=YY
        B=Y
        U=J
9  CONTINUE
   IF(BB.LT.AA) GOTO 10
   IT=IT+1
   GOTO 6
10 IT=0
   D=D-AA+BB
   E(R)=A
   E(U)=B
   Q(R)=Q(R)-1
   Q(U)=Q(U)+1
   P(I)=U
   IF(IDR.NE.0) WRITE(KO,11) I,D,(P(V),V=1,M)
11 FORMAT(1X,I4,F12.2,23I3/(17X,22I3))
   GOTO 6
   END
```

Fig. U 8

Algorithm CLUDIA (Fig. U8) implements the step-by-step reduction of Z_1 in accordance with (3.2.32) or (3.2.33) depending respectively on whether T(I,K) = $d^2(x_i, x_k)$ or T(I,K) = $d(x_i, x_k)$ is input. Exactly as in KMEANS, an initial partitioning of p_i ($i = 1, \ldots, m$) must be assumed, which is inserted into the main program H7 unless it had been read in earlier in accordance with (3.2.10). Furthermore, in H7, by way of contrast to H6 and H5, the upper triangular matrix T(J,K) (J = 1, . . . ,M − 1; K = J + 1, . . . ,M) is read in, the elements of which are unchanged for IT = 0, and those for IT≠0 are provided to CLUDIA in squared form. If squared Euclidean distances are contained in the T matrix, and the same initial partition is used, CLUDIA returns the same result as KMEANS, for which reason we are not repeating any of the examples which have been considered already. This will also be true, for L_1 – distances, for CLUDIA and EMEANS from section 3.5. CLUDIA can, however, be used for arbitrary distance values that are not necessarily calculated from metric data (see Langenmayr and Späth 1976); we shall introduce examples of this in section 4.2 (see also Späth 1977c, 1977d, Braun 1978).

Fig. H 7

```
C
C     CLUSTERING VON SYMMETRISCHEN MATRIZEN
C
      DIMENSION T(60,60),P(60),E(10)
      INTEGER P
      KI=5
      KO=6
    1 READ(KI,2) M,N1,N2,IDR,IP,IT
    2 FORMAT(16I5)
      IF(M.LE.1.OR.M.GT.60.OR.N1.GT.M.OR.N1.LT.1
     *   .OR.N2.LT.N1.OR.N2.GT.10) STOP
      WRITE(KO,3) M,N1,N2,IDR,IP,IT
    3 FORMAT('1','  M=',I2,'  N1=',I2,'  N2=',I2,
     *       '  IDR=',I1,'  IP=',I1,'  IT=',I1)
      WRITE(KO,4)
    4 FORMAT('0')
      M1=M-1
      DO 6 J=1,M1
            J1=J+1
            READ(KI,7) (T(J,K),K=J1,M)
            DO 5 K=J1,M
                  A=T(J,K)
                  IF(IT.NE.0) A=A*A
                  T(J,K)=A
                  T(K,J)=A
    5       CONTINUE
    6 CONTINUE
    7 FORMAT(16F5.0)
      DO 8 J=1,M
            T(J,J)=0.
            WRITE(KO,9) (T(J,K),K=1,M)
    8 CONTINUE
```

```
    9 FORMAT(1X,10F8.1)
      DO 14 N=N1,N2
            WRITE(KO,4)
            IF(IP.NE.0) GOTO 11
            K=0
            DO 10 I=1,M
                  K=K+1
                  IF(K.GT.N) K=K-N
                  P(I)=K
   10       CONTINUE
            GOTO 12
   11       READ(KI,2)    (P(I),I=1,M)
   12       WRITE(KO,13) (P(I),I=1,M)
   13       FORMAT(1X,10I7)
            CALL CLUDIA (M,T,P,N,E,D,IDR)
            WRITE(KO,4)
            WRITE(KO,13) (P(I),I=1,M)
            WRITE(KO,4)
            WRITE(KO,15) (E(J),J=1,N),D
   14 CONTINUE
   15 FORMAT(1X,10F7.1)
      GOTO 1
      END
```

Fig. H 7

It is clear too that CLUDIA is more efficient in terms of calculation time than KMEANS, since the distances are calculated once and for all before executing the passes through the objects. For metric data, and also for (non-negative) distance values obtained in other ways, this efficiency is limited by the number of objects and the computer storage available. For m objects which are measured over l variables, KMEANS essentially needs $m.l$ storage places, while CLUDIA is not affected by l but requires at least $m(m - 1)/2$ storage positions for the distance matrix (for one-dimensional storage of the upper triangular matrix, which we have not used in CLUDIA) (see Braun 1978). Therefore CLUDIA is limited to about $m \leqslant 200$ objects, whereas KMEANS given similar storage can handle ($m.l \leqslant 20{,}000$) e.g. 2000 objects over 10 variables. In addition it is not possible to process distances sequentially using CLUDIA, as can be done with KMEANS, and hence it is not so easy to exploit external storage.

The same considerations would apply if CLUDIA were modified for the objective functions (3.2.34) and (3.2.35).

Similar considerations generally apply in the application of distance matices in cluster algorithms, which is often decisive for nominal, ordinal or mixed data. One can certainly use external storage for the matrix; however, this demands a cumbersome data organisation and one which essentially weakens the effectiveness of the algorithm. By appropriate sorting (see Anderberg 1973), improvements can be obtained. Particular algorithms are suited to distance matrices – as we shall see in section 5.2 – in that each distance $d(x_i, x_k)$ is used just once in the algorithm and for this reason only needs to be calculated at that point, which avoids storing the whole matrix. In general, the more satisfactory algorithms do not have this characteristic.

3.3 INVARIANT CRITERIA

In the last section we have seen that the objective function from (3.2.5) and its re-formulation (3.2.26) for $l > 1$ is invariant only with respect to translations and rotations of the row vectors $x_i = x_{i.}^T$ of the (metric) data matrix, and hence the question of an appropriate scaling emerges.

In this section we examine objective functions which are invariant with respect to all non-singular transformations, and thus in particular are invariant with respect to scale. We assume a metric data matrix is a prerequisite here. We introduced the Mahalanobis distance d_s, which is invariant with respect to non-singular transformations, in (2.1.27). One way to derive invariant objective functions follows from (3.2.26) and modified criteria like

$$\tilde{Z}_1(C_1, \dots, C_n) = \sum_{j=1}^{n} \frac{1}{m_j} \sum_{i \in C_j} \sum_{\substack{k \in C_j \\ k > i}} d_S^2(x_i, x_k) \tag{3.3.1}$$

or

$$\tilde{Z}_1(C_1, \dots, C_n) = \sum_{j=1}^{n} \frac{1}{m_j} \sum_{i \in C_j} \sum_{\substack{k \in C_j \\ k > i}} d_S(x_i, x_k) \tag{3.3.2}$$

or

$$\tilde{Z}_1(C_1, \dots, C_n) = \sum_{j=1}^{n} \sum_{i \in C_j} \sum_{\substack{k \in C_j \\ k > i}} d_S(x_i, x_k). \tag{3.3.3}$$

The objective functions (3.3.1) and (3.3.2) could be directly used by the CLUDIA subroutine, in which the Mahalanobis distance is presumed to be the distance matrix. For (3.3.3) minor modifications would have to be made.

In (Fukunaga and Koontz 1970, Howard 1966) an objection is advanced to the method applied in CLUDIA – which can also be adopted with the KMEANS principle (Duda and Hart 1973) – namely, using in d_s a covariance variable matrix consisting of all the row vectors of the data matrix. Their grounds are that a unique covariance matrix should be used for each cluster, made up of the x_i vectors belonging to that cluster, although of course the composition of the clusters is not known beforehand. For this reason, it might in practice be preferable to use, instead of (3.3.1), the following objective function, which appears in a similar form to (3.2.5) in (Diday and Govaert 1974):

$$Z_1'(C_1,\ldots,C_n) = \sum_{j=1}^{n} \frac{1}{m_j} \sum_{i\in C_j} \sum_{\substack{k\in C_j \\ k>i}} d_{S_j}^2(x_i, x_k) \tag{3.3.4}$$

where

$$S_j = (s_{ik}^{(j)}) = \frac{1}{m_j} \sum_{q\in C_j} (x_{qi} - \bar{x}_{.i})(x_{qk} - \bar{x}_{.k}). \tag{3.3.5}$$

This objective function and the variants corresponding to (3.3.2) and (3.3.3) are likewise invariant with respect to non-singular transformations of the vectors x_i. Späth and Müller (1979) have given an exchange method for an objective function like (3.3.4) in which the matrices (3.3.5) are suitably normalised.

Analogous to the process for (3.1.7) a relationship corresponding to (3.2.4) for variance matrices can be derived:

$$\sum_{i=1}^{m} (x_i - \bar{x})(x_i - \bar{x})^T \tag{3.3.6}$$

$$= \sum_{j=1}^{n} \sum_{i\in C_j} (x_i - \bar{x}_j)(x_i - \bar{x}_j)^T + \sum_{j=1}^{n} m_j (\bar{x}_j - \bar{x})(\bar{x}_j - \bar{x})^T.$$

We write as an abbreviation of (3.3.6):

$$T = W + B. \tag{3.3.7}$$

Here T, W and B are l by l matrices, which represent respectively: the total variance in the data-matrix, the sum of the variances within the clusters, and those between the clusters. Normally T is positive definite for $m-1 \geqslant l$ (see Kuganawa and Koontz 1970) and W is positive definite for $m - n \geqslant l$ (see Friedman and Rubin 1967, Duda and Hart 1973).

The relationship between (3.3.7) and (3.2.4) is obtained by forming the trace – that is, summing the diagonal elements – so that Z_1 from (3.2.5) can therefore be written as

$$Z_1(C_1,\ldots,C_n) = \text{trace } W \to \min \tag{3.3.8}$$

and Z_2 from (3.2.6) as

$$Z_2(C_1,\ldots,C_n) = \text{trace } B \to \max \tag{3.3.9}$$

(The trace of a square matrix is the sum of its diagonal elements and also equals the sum of its eigenvalues.) On the other hand one must now ask how one gets a criterion which corresponds to (3.3.8) but which remains invariant.

For $l = 1$ T, W and B are scalars, and the criteria

$$\frac{B}{W} \longrightarrow \max$$

$$\frac{T}{W} \longrightarrow \max$$

$$W \longrightarrow \max$$

are identical (see Engelman and Hartigan 1969). Thus when $l = 1$ the objective function does not alter if one applies the trace on the left-hand side.

For $l \geqslant 1$ the corresponding objective functions become:

$$\text{trace } (W^{-1} B) \rightarrow \max \tag{3.3.10}$$

$$\text{trace } (W^{-1} T) \rightarrow \max \tag{3.3.11}$$

$$\text{trace } (W) \quad \rightarrow \min \tag{3.3.12}$$

Another scalar mapping, which can be intepreted as a measurement of volume, is the determinant, and one can analogously form the objective functions:

$$\det (W^{-1} B) \rightarrow \max \tag{3.3.13}$$

$$\det (W^{-1} T) \rightarrow \max \tag{3.3.14}$$

and

$$\det (W) \quad \rightarrow \min \tag{3.3.15}$$

From all these objective functions, let us set aside (3.3.12) since it is already known, and also (3.3.14) and (3.3.15) which are identical because det T is constant. We wish to select those that can be represented as functions of the eigenvalues λ_k $(k = 1, \ldots, l)$ of the matrix $W^{-1} B$. These eigenvalues possess the required invariance properties (see Fukunaga and Koontz 1970, Duda and Hart 1973). This can readily be verified, as follows from (2.1.27).

Now we have

$$\text{trace } (W^{-1} B) = \sum_{k=1}^{l} \lambda_k$$

$$\text{trace } (W^{-1} T) = \sum_{k=1}^{l} (1 + \lambda_k)$$

$$\det\ (W^{-1} B) = \prod_{k=1}^{l} \lambda_k$$

$$\det\ (W^{-1} T) = \prod_{k=1}^{l} (1 + \lambda_k).$$

Here again the first two objective functions are identical as far as maximisation is concerned. The last can be written

$$\det W = \frac{\det T}{\prod_{k=1}^{l} (1 + \lambda_k)}$$

Accordingly the following remain as invariant objective functions:

$$Z_3 (C_1, \ldots, C_n) = \det W \rightarrow \min \tag{3.3.16}$$

$$Z_4 (C_1, \ldots, C_n) = \text{trace } (W^{-1} T) \rightarrow \max \tag{3.3.17}$$

$$Z_5 (C_1, \ldots, C_n) = \det (W^{-1} B) \rightarrow \max \tag{3.3.18}$$

In addition (Anderberg 1973) gives

$$Z_6 (C_1, \ldots, C_n) = \max_k \lambda_k \rightarrow \min$$

which is also meaningful since $\lambda_k \geqslant 0$ $(k = 1, \ldots, l)$. A Fortran program is described in (McRae 1973), which tries to minimise according to the KMEANS principle, a choice of one of the functions Z_1, Z_3, Z_4 and Z_5, making use, also at choice, of the Euclidean distance, a weighted Euclidean distance or the Mahalanobis distance.

In what follows we shall consider only the objective function Z_3, following the recommendation of (Friedman and Rubin 1967, Marriot 1971, Scott and Symons 1971). This is easily constructed, and we develop the formulae necessary to apply the KMEANS principle to it. From the matrix definition of the dyadic product

$$E_q = \sum_{i \in C_q} (x_i - \bar{x}_q)(x_i - \bar{x}_q)^T \tag{3.3.19}$$

we obtain, for $C_p = C_j - \{k\}$, corresponding to (3.2.17),

$$E_p = E_j - \frac{m_j}{m_j - 1} (\bar{x}_j - x_k)(\bar{x}_j - x_k)^T \tag{3.3.10}$$

and, for $C_p = C_j \cup \{k\}$, corresponding to (3.2.21),

$$E_p = E_j + \frac{m_j}{m_j + 1} (\bar{x}_j - x_k)(\bar{x}_j - x_k)^T. \tag{3.3.21}$$

Now all the points x_i, where $i \in C_r$, are again considered in turn, and the point x_i is tentatively transferred into every cluster with serial number $j \neq r$. If, for at least one j,

$$W_j^* = W - \frac{m_r}{m_r - 1}(\bar{x}_r - x_i)(\bar{x}_r - x_i)^T + \frac{m_j}{m_j + 1}(\bar{x}_j - x_i)(\bar{x}_j - x_i)^T \quad (3.3.22)$$

is such that

$$\det W_j^* < \det W, \quad (3.3.23)$$

then x_i is transferred to that cluster ν, for which the left-hand side of (3.2.23) becomes smallest. There is an implied condition, of course, that W is positive definite for $m - n \geqslant l$. Corresponding algorithms for the objective functions (3.3.17) and (3.3.18) are more complicated since the $(W_j^*)^{-1}$ has to be formed instead of det W_j^*.

This algorithm is implemented in subroutine WMEANS (Fig. U9). It calls the auxiliary routines DYAD (Fig. U10), to calculate the dyadic product $z\,z^T$ of a column vector z, and DET (Fig. U11) (see Forsythe and Moler 1971), to calculate the determinant of a matrix. Neither the fact that W is symmetric nor updating is the determinants has been exploited. All floating point quantities, other than the data matrix X(M,L), are held in double precision, since this reduces the loss of accuracy which would otherwise occur in the calculation, particularly of (3.3.22). WMEANS, in fact, seems empirically to be more liable to rounding errors than KMEANS. The main program H8 is again almost identical to H5 and H6. The input data, in particular, is expected in the same form.

Fig. U 9

```
      SUBROUTINE WMEANS (M,L,X,P,N,S,DETM,IDR)
C
C     LET DETM INDICATE THE DETERMINANT OF THE SUM, FORMED
C     OVER THE N CLUSTERS, OF THE DYADIC PRODUCTS - WHICH ARE
C     MATRICES - OF THE DIFFERENCES OF THE CLUSTER ELEMENTS
C     FROM THEIR CENTROID.
C
C     AS IN THE KMEANS SUBROUTINE, ALL BUT TWO OF WHOSE
C     PARAMETERS ALSO APPEAR HERE, THE DETERMINANT IS MADE AS
C     SMALL AS POSSIBLE BY REPEATED EXCHANGE OF CLUSTER MEMBERS.
C
      DOUBLE PRECISION S,F,DM,Z,AM,H,G,BM,DN,DV,A,B
      INTEGER   P,Q,R,U,V,W
C     DIMENSION X(M,L),   P(M),   S(N,L),   Q(N)
      DIMENSION X(100,12),P(100),S(10,12),Q(10)
C     DIMENSION Z(L), DM(L,L),   DN(L,L),
      DIMENSION Z(12),DM(12,12),DN(12,12)
C     DIMENSION DV(L,L),   AM(L,L),   BM(L,L)
      DIMENSION DV(12,12),AM(12,12),BM(12,12)
      KO=6
      DO 2 J=1,N
         Q(J)=0
         DO 1 K=1,L
            S(J,K)=0.
```

```
 1         CONTINUE
 2  CONTINUE
    DO 4  I=1,M
          R=P(I)
          IF(R.LT.1.OR.R.GT.N) RETURN
          Q(R)=Q(R)+1
          DO  3 K=1,L
                S(R,K)=S(R,K)+DBLE(X(I,K))
 3        CONTINUE
 4  CONTINUE
    DO 6 J=1,N
          R=Q(J)
          IF(R.EQ.0) RETURN
          F=1./DFLOAT(R)
          DO  5 K=1,L
                S(J,K)=S(J,K)*F
 5        CONTINUE
 6  CONTINUE
    DO 8  K=1,L
          DO 7 KK=K,L
                DM(K,KK)=0.
 7        CONTINUE
 8  CONTINUE
    DO 12 I=1,M
           R=P(I)
           DO 9 K=1,L
                Z(K)=DBLE(X(I,K))-S(R,K)
 9         CONTINUE
           CALL DYAD (L,Z,AM)
           DO 11 K=1,L
                  DO 10 KK=K,L
                         DM(K,KK)=DM(K,KK)+AM(K,KK)
10                CONTINUE
11         CONTINUE
12  CONTINUE
    CALL DET (L,DM,DETM)
    I=0
    IT=0
13  I=I+1
    IF(I.GT.M) I=I-M
    IF(IT.EQ.M) RETURN
    R=P(I)
    U=Q(R)
    IF(U.LE.1) GOTO 13
    H=DFLOAT(U)
    F=H/(H-1.)
    DO 14 K=1,L
          Z(K)=DBLE(X(I,K))-S(R,K)
14  CONTINUE
    CALL DYAD (L,Z,AM)
    DETV=1.E30
    DO 20 J=1,N
           IF(R.EQ.J) GOTO 20
           U=Q(J)
           G=DFLOAT(U)
           G=G/(G+1.)
           DO 15 K=1,L
                  Z(K)=DBLE(X(I,K))-S(J,K)
15         CONTINUE
           CALL DYAD (L,Z,BM)
           DO 17 K=1,L
                  DO 16 KK=K,L
                         DM(K,KK)=DM(K,KK)-F*AM(K,KK)
   *                                     +G*BM(K,KK)
```

```
16                  CONTINUE
17            CCNTINUE
              CALL DET (L,DN,DETN)
              IF(DETN.GT.DETV) GOTO 20
              DETV=DETN
              DO 19 K=1,L
                    DO 18 KK=K,L
                          DV(K,KK)=DN(K,KK)
18                  CONTINUE
19            CONTINUE
              V=J
              W=U
20      CONTINUE
        IF(DETV.LT.DETM) GOTO 21
        IT=IT+1
        GOTO 13
21      IT=0
        DETM=DETV
        DO 23 K=1,L
              DO 22 KK=K,L
                    DM(K,KK)=DV(K,KK)
22            CCNTINUE
23      CONTINUE
        A=1./(H-1.)
        G=DFLOAT(W)
        B=1./(G+1.)
        DO 24 K=1,L
              F=DBLE(X(I,K))
              S(R,K)=(H*S(R,K)-F)*A
              S(V,K)=(G*S(V,K)+F)*B
24      CCNTINUF
        P(I)=V
        Q(R)=Q(R)-1
        Q(V)=Q(V)+1
        IF(IOR.EQ.1) WRITE(KO,25) I,DETM,(P(U),U=1,M)
25      FORMAT(1X,I4,E15.5,23I3/(20X,23I3))
        GOTO 13
        END
```

Fig. U 10

```
        SUBROUTINE DYAD (N,Z,A)
C
C       FORMATION OF THE DYADIC PRODUCT (THE UPPER TRIANGULAR PART
C       OF THE MATRIX ONLY, INCLUDING THE DIAGONAL) OF A VECTOR Z
C       OF LENGTH N,
C
        DOUBLE PRECISION Z,A,T
C       DIMENSION Z(N), A(N,N)
        DIMENSION Z(12),A(12,12)
        DO 2 I=1,N
              T=Z(I)
              DO 1 K=I,N
                    A(I,K)=T*Z(K)
    1         CONTINUE
    2   CCNTINUE
        RETURN
        END
```

```
      SUBROUTINE DET (N,A,DETA)
C
C     CALCULATION OF THE DETERMINANT OF A SYMMETRICAL
C     N BY N MATRIX, A.    ONLY THE UPPER TRIANGULAR
C     PART OF A, INCLUDING THE DIAGONAL, HAS TO BE GIVEN.
C
      DOUBLE PRECISION A,B,T
C     DIMENSION A(N,N),  B(N,N),   IP(N)
      DIMENSION A(12,12),B(12,12),IP(12)
      DO 2 K=1,N
           DO 1 I=K,N
                T=A(K,I)
                B(I,K)=T
                B(K,I)=T
    1      CONTINUE
    2 CONTINUE
      IP(N)=1
      DO 8 K=1,N
           IF(K.EQ.N) GOTO 7
           K1=K+1
           M=K
           DO 3 I=K1,N
                IF(DABS(B(I,K)).GT.DABS(B(M,K))) M=I
    3      CONTINUE
           IP(K)=M
           IF(M.NE.K) IP(N)= - IP(N)
           T=B(M,K)
           B(M,K)=B(K,K)
           B(K,K)=T
           IF(T.EQ.0.) GOTO 7
           T=1./T
           DO 4 I=K1,N
                B(I,K)= - B(I,K)*T
    4      CONTINUE
           DO 6 J=K1,N
                T=B(M,J)
                B(M,J)=B(K,J)
                B(K,J)=T
                IF(T.EQ.0.) GOTO 6
                DO 5 I=K1,N
                     B(I,J)=B(I,J)+B(I,K)*T
    5           CONTINUE
    6      CONTINUE
    7      IF(B(K,K).EQ.0.) IP(N)=0
    8 CONTINUE
      T=DFLOAT(IP(N))
      DO 9 K=1,N
           T=T*B(K,K)
    9 CONTINUE
      DETA=T
      RETURN
      END
```

Fig. U 11

```
C
C     WMEANS-CLUSTERING WITH A GIVEN INITIAL PARTITION
C
      DOUBLE PRECISION S
      INTEGER P
      DIMENSION X(100,12),P(100),S(10,12)
      KI=5
      KO=6
    1 READ(KI,2) M,L,N1,N2,IDR,IP
    2 FORMAT(16I5)
      IF(M.LE.1.OR.M.GT.100.OR.L.LT.1.OR.L.GT.12.OR.
     *   N1.GT.M.OR.N1.LT.1.OR.N2.LT.N1.OR.N2.GT.10) STOP
      WRITE(KO,3) M,L,N1,N2,IDR,IP
    3 FORMAT('1','  M=',I3,'  L=',I2,'  N1=',I2,'  N2=',
     *       I2,'  IDR=',I1,'  IP=',I1)
      WRITE(KO,4)
    4 FORMAT('0')
      DO 5 K=1,L
         READ(KI,6)  (X(I,K),I=1,M)
         WRITE(KO,7) (X(I,K),I=1,M)
    5 CONTINUE
    6 FORMAT(16F5.0)
    7 FORMAT(1X,10F7.1)
      DO 13 N=N1,N2
            WRITE(KO,4)
            IF(IP.NE.0) GOTO 9
            K=0
            DO 8 I=1,M
                 K=K+1
                 IF(K.GT.N) K=K-N
                 P(I)=K
    8       CONTINUE
            GOTO 10
    9       READ(KI,2)    (P(I),I=1,M)
   10       WRITE(KO,11) (P(I),I=1,M)
   11       FORMAT(1X,10I7)
            CALL WMEANS (M,L,X,P,N,S,DET,IDR)
            WRITE(KO,4)
            WRITE(KO,11) (P(I),I=1,M)
            WRITE(KO,4)
            DO 12 K=1,L
                  WRITE(KO,7) (S(J,K),J=1,N)
   12       CONTINUE
            WRITE(KO,4)
            WRITE(KO,14) DET
   13 CONTINUE
   14 FORMAT(1X,'DET =',E15.5)
      GOTO 1
      END
```

Fig. H 8

Nearly the same results are obtained for problems of the kind posed by the 22 and 59 towns (Table T1 and Fig. E6.2 respectively), with a similar final partitioning, whether one uses WMEANS or KMEANS. The discrepancies in assignment are so minute that it is not possible without computation to decide if one of the two allocations is better than the other with respect to the two objective functions. Similarly the number of the passes required, given for $m = 22$ in Table T10 and for $m = 59$ in Table T11, are nearly comparable with those in T2 and T3:

Table T10

n	det W	1	2	3	4	5
2	.994E9	11	1	1	2	
3	.277E9	11				
4	.149E9	14	1			
5	.650E8	15	2	2	1	1

Table T11

n	det W	1	2	3	4	5	6	7
2	.331E10	26	3					
3	.131E10	34	3					
4	.503E9	41	12	1				
5	.346E9	43	15	6	3			
6	.259E9	48	13	3	1	2	1	2
7	.162E9	41	13	2	2	6	2	
8	.114E9	45	14	4	5	2		
9	.913E8	50	19	5	2	1	3	3
10	.698E8	50	18	10	9			

For examples like E6.15, where one can now apply scaled or non-scaled data, the differences are significant (see Everitt 1974). The results selected should always be those most amenable to interpretation.

Fig. B22 shows the results of KMEANS and WMEANS for $m = 6$ points in a plane ($l = 2$) with various scale transformations, with $n = 2$ and $n = 3$, starting from the initial allocations shown in the bottom right diagram. WMEANS of course gives the same final partitioning whatever the scale transformation. The results of KMEANS are seen to be different and even more significant – in a purely geometric sense – for $n = 3$ and for the system of coordinates $(x, y/5)$, although this may be due to the initial assignment chosen.

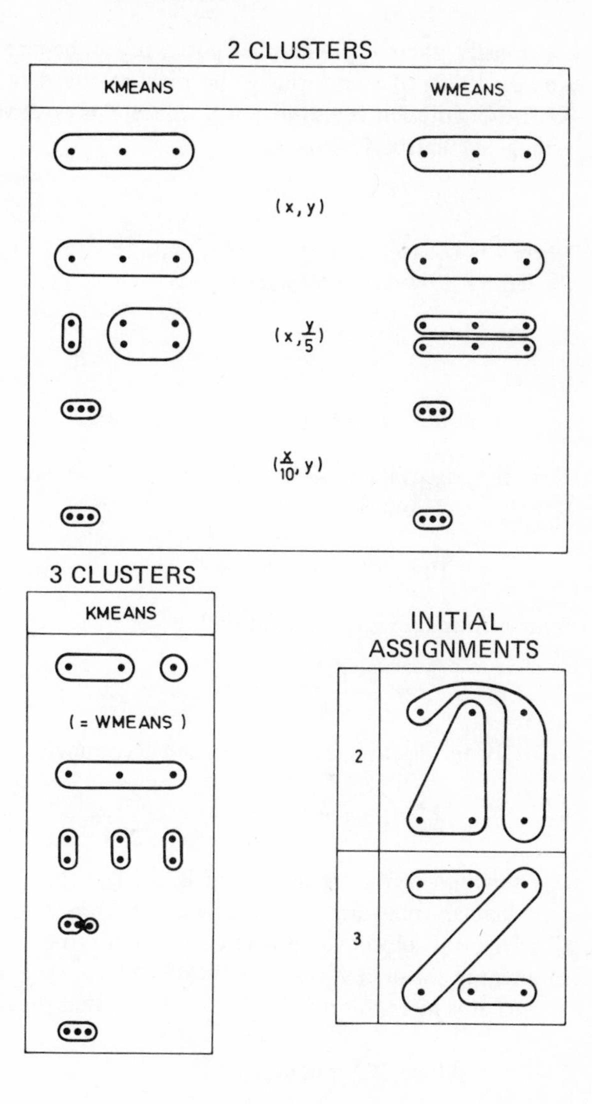

Fig. B 22

For cases where *m, n* and *l* are of the orders of magnitude that have occurred in the examples we have looked at so far, WMEANS has been shown empirically to require ten to twenty times as much calculation as KMEANS. For this reason WMEANS should be used only if it is very uncertain what scaling should be chosen for KMEANS; and, in the event that WMEANS is used, the final partition produced by KMEANS, in whatever scaling there is, can be used as an initial partition for WMEANS. (see also Späth 1977b). If the sealing is a hard problem, we recommend the subroutine AKMEAN (Späth and Müller 1979).

Finally, there is yet another possibility to be mentioned (see Fukunaga and Koontz 1970) of transforming the rows of the data matrix (1.1) so that the KMEANS criterion is also invariant. Since T is positive definite, there exists an orthogonal matrix A with

$$ATA^T = I \tag{3.3.24}$$

where I is the identity matrix. If we multiply (3.3.7) on the right by A^T and on the left by A, we get accordingly

$$I = W^* + B^* \tag{3.3.25}$$

where

$$W^* = AWA^T, \quad B^* = ABA^T. \tag{3.3.26}$$

Now the eigenvalues of $W^{-1}B$ and $(W^*)^{-1}B^*$ are identical and equal to λ_k $(k = 1, \ldots, l)$. Since

$$(W^*)^{-1} = I + (W^*)^{-1}B^*$$

follows from (3.3.25), the relationship

$$\mu_k = \frac{1}{1+\lambda_k} \quad (k=1,\ldots,l) \tag{3.3.27}$$

holds for the eigenvalues μ_k of W^*, and accordingly

$$Z_1^* = \text{trace } W^* = \sum_{k=1}^{l} \mu_k = \sum_{k=1}^{l} \frac{1}{1+\lambda_k} \tag{3.3.28}$$

is a function of the eigenvalues of $W^{-1}B$ and thus invariant with respect to all non-singular transformations. If one therefore determines the matrix A from (3.3.24), the objective function (3.2.5) for the transformed vectors $Ax_{i.}^T$ is invariant in the desired sense, and KMEANS can be applied.

What is more, for $n = 2$ it can be shown that the objective functions

$$\text{trace } W^* \to \min$$

$$\det W \to \min$$

$$\text{trace } W^{-1}B \to \max$$

are equivalent (see Fukunaga and Koontz 1970), i.e. possess the same optimum partitions. This circumstance can still be of importance in section 5.2, as the set of objects is successively split up into two clusters. Generally for $n > 2$ it is still the case that the closer to zero λ_k are, the more nearly equivalent are the objective functions.

The underlying problem with all objective functions – and the number of them which could be considered infinite, because any function of the eigenvalues $\lambda_1, \ldots, \lambda_l$ of $W^{-1} B$ is suitable – is that no particular one is obviously better than all the others. (The choice of (3.3.16) for a program was made because of its suitability for computation.) In the long run this is just as unsatisfactory as the defective scale invariance property of Z_1. Sadly, it is as always the algorithm which yields the most easily interpreted results which is the one to be recommended.

3.4 THE MULTIPLE LOCATION-ALLOCATION PROBLEM

A problem which arises in economics is, how to determine the location of one or more supply points ($n \geqslant 1$) for a number of locations m, in such a way as to minimise the (possibly weighted) sum of the distances between the supply points and their associated locations. Transport costs are normally proportional to distance. Thus the geographical Cartesian coordinates of the locations give a data matrix (1.1). Normally only the case $l = 2$ is discussed (see Bloech 1970). We shall permit $l \geqslant 2$ here.

The problem is described by the objective function Z_1 of (3.2.5), if on the right-hand side the power 2 is omitted. We now explicitly permit weights $w_i > 0$ $(i = 1, \ldots, m)$ for the objects x_i. Thus the objective function to be minimised is

$$\widetilde{Z}_1 (C_1, \ldots, C_n) = \sum_{j=1}^{n} \sum_{i \in C_j} w_i \, \|x_i - \widetilde{x}_j\|, \tag{3.4.1}$$

in which the vectors $\widetilde{x}_j$ now represent the optimal supply centres we are trying to find, rather than centroid coordinates.

Let us first consider the case of a single centre ($n = 1$), and hence of one cluster $C_1 = \{1, \ldots, m\}$. The function to be minimised with respect to $\widetilde{x} = \widetilde{x}_1$ then becomes

$$\begin{aligned} F(\widetilde{x}) = Z_1(C_1) &= \sum_{i=1}^{m} w_i \, \|x_i - \widetilde{x}\| \\ &= \sum_{i=1}^{m} w_i \sqrt{\sum_{k=1}^{l} (x_{ik} - \widetilde{x}_k)^2}. \end{aligned} \tag{3.4.2}$$

Since $F = F(\widetilde{x})$ is a convex function, the minimum sought is unique and lies in the convex hull of the vectors x_i, but can be determined exactly only in very special cases (see Bloech 1970). In general one will try to find an iterative solution of the necessary and also (because of the convexity) sufficient conditions for a minimum, which are

$$\frac{\partial F}{\partial \tilde{x}_k} = 2 \sum_{i=1}^{m} \frac{w_i (x_{ik} - \tilde{x}_k)}{\|x_i - \tilde{x}\|} = 0, \tag{3.4.3}$$

Newton's method always converges – that is, converges for all initial values (see Ortega and Rheinbolt 1970). Simpler, in terms of the computation required, is the fixed point method

$$\tilde{x}_k^{(t+1)} = \frac{\displaystyle\sum_{i=1}^{m} \frac{w_i \tilde{x}_{ik}^{(t)}}{\|x_i - \tilde{x}_k^{(t)}\|}}{\displaystyle\sum_{i=1}^{m} \frac{w_i}{\|x_i - \tilde{x}_k^{(t)}\|}} \qquad (k=1,\dots,l). \tag{3.4.4}$$

which is easy to derive from (3.4.2). In this, t is an iteration index. For the initial value ($t = 0$) the coordinates of the centroid

$$\tilde{x}_k^{(0)} = \frac{\displaystyle\sum_{i=1}^{m} w_i x_{ik}}{\displaystyle\sum_{i=1}^{m} w_i} \tag{3.4.5}$$

can be chosen. The convergence of method (3.4.4) has remained unproved for a long time (see Bloech 1970, Cooper 1963). (Kuhn 1973) was able to show incontestably that the method converges apart from certain exceptional cases which seldom arise in practice because of rounding errors.

This method is implemented in the subroutine STANDN in Fig. U12. With EPS = 0.001 as the value of the accuracy parameter, between ten and, at the most, twenty iterations can be expected using (3.4.5) for the initial value.

Fig. U 12

```
      SUBROUTINE STANDN (M,L,X,W,S,EPS,ITMAX,IS,F)
C
C     F(S) = SUM(W(I)*SQRT(SUM((S(K)-X(I,K))**2))
C     IS MINIMISED.   THIS CORRESPONDS TO THE SOLUTION OF THE
C     SINGLE LOCATION PROBLEM IN L DIMENSIONS.
C
C     EPS IS THE VALUE USED IN AN ACCURACY TEST.
C
C     ITMAX IS THE MAXIMUM NUMBER OF ITERATIONS TO BE CARRIED
C     OUT.   (ON RETURN, ITMAX IS REPLACED BY THE ACTUAL
C     NUMBER OF ITERATIONS.)
C
C     IF THE 'IS' PARAMETER = 0, THE CENTROID IS TAKEN AS THE
C     INITIAL LOCATION OF THE VALUE.   FOR 'IS' NON-ZERO, THE
C     INITIAL COORDINATES ARE GIVEN IN S.
```

```
C
C      THE COMBINATION OF 'IS' NON-ZERO AND ITMAX = 1 CAN BE
C      USED TO TABULATE F.
C
C      IF, DURING THE ITERATION, THE CENTRE CONVERGES TO A
C      GIVEN POINT, THE 'IS' PARAMETER IS SET TO -1 AND
C      THE PROCEDURE TERMINATED.
C
C
       DIMENSION X(M,L),    W(M),   S(L), T(L)
       DIMENSION X(100,12),W(100),S(12),T(12)
       LOGICAL PTRUE
       IT=0
       IF(IS.NE.0) GOTO 5
       DO 1 K=1,L
            S(K)=0.
     1 CONTINUE
       P=0.
       DO 3 I=1,M
            V=W(I)
            P=P+V
            DO 2 K=1,L
                 S(K)=S(K)+V*X(I,K)
     2      CONTINUE
     3 CONTINUE
       P=1./P
       DO 4 K=1,L
            S(K)=S(K)*P
     4 CONTINUE
       F=0.
       IF(M.EQ.1) GOTO 11
     5 IT=IT+1
       IF(IT.GT.ITMAX) RETURN
       DO 6 K=1,L
            T(K)=0.
     6 CONTINUE
       F=0.
       Z=0.
       DO 9 I=1,M
            V=W(I)
            P=0.
            DO 7 K=1,L
                 H=S(K)-X(I,K)
                 P=P+H*H
     7      CONTINUE
            PTRUE=P.LT.1.E-10
            IF(PTRUE) GOTO 9
            P=SQRT(P)
            F=F+V*P
            P=V/P
            Z=Z+P
            DO 8 K=1,L
                 T(K)=T(K)+P*X(I,K)
     8      CONTINUE
     9 CONTINUE
       IF(PTRUE) GOTO 12
       P=0.
       V=0.
       Z=1./Z
       DO 10 K=1,L
            Y=T(K)*Z
            V=V+ABS(Y)
            P=P+ABS(Y-S(K))
            S(K)=Y
```

```
10 CONTINUE
   IF(P.GE.EPS*V) GOTO 5
11 ITMAX=IT
   RETURN
12 IS= - 1
   ITMAX=IT
   RETURN
   END
```

Fig. U 12

Let us now examine the minimisation of the objective function $\widetilde{Z}_1$ for $n > 1$ from (3.4.1) using the KMEANS principle. Starting from the initial partition, the locations are assigned in turn to other cluster centres if this produces a reduction in the objective function. Since there is no easy transformation of the centres as there was earlier with the centroids, the centres for a reduced cluster and for all clusters concerned must be explicitly recalculated by means of STANDN at each iteration, which is cumbersome. As always, a location is reassigned to that cluster producing the greatest reduction in the objective function $\widetilde{Z}_1$.

The routine corresponding to this process is called CLUSTA and is given in Fig. U13. If STANDN 'fails' in the sense that the iteration cannot be carried out further because a centre approaches a location so closely that the denominator in (3.4.4) oscillates about zero, it is assumed that the centre is placed at this location. Overall, the organisation of CLUSTA is somewhat more complicated than that of KMEANS or WMEANS. The main program H9 is, however, almost identical to H5, H6 and H8. The initial partitions corresponding to (3.2.10) are also produced as usual.

Fig. U 13

```
      SUBROUTINE CLUSTA (M,L,X,W,P,N,S,E,D,IDR)
C
C     THE INITIAL ASSIGNMENT OF THE VECTORS X(M,*) TO N CLUSTERS
C     (LOCATION CENTRES) IS GIVEN BY THE ARRAY P, WHERE P(I) IS
C     THE CLUSTER NUMBER OF THE I-TH VECTOR, THUS EACH P(I) MUST
C     BE SUCH THAT 1 <= P(I) <= N AND FOR EACH J=1,...,N AT LEAST
C     ONE I WITH P(I) = J MUST EXIST.
C
C     LET S(J,L) INDICATE THAT VECTOR FOR WHICH THE SUM OF THE
C     DISTANCES BETWEEN IT AND ITS CLUSTER ELEMENTS, WEIGHTED
C     BY W(I), I.E.
```

```
C
C            E(J) =  SUM  (W(I)*SQRT(SUM((S(J,K)-X(I,K))**2)))
C                   P(I)=J             K
C      IS MINIMISED.
C
C      THE SUBROUTINE MINIMISES D = SUM(E(J)) AS FAR AS POSSIBLE
C      BY REPEATED EXCHANGE OF CLUSTER MEMBERS.   THE P(I) ARE
C      CORRESPONDINGLY MODIFIED BUT THE NUMBER OF CLUSTERS IS NOT.
C
C      THE SUBROUTINE RETURNS THE VALUES AT THE FINAL
C      CONFIGURATION OF THE LOCATIONS S(N,L), THE
C      SUMS E(N), AND D.
C
C      IF IDR = 1, THE CURRENT VALUES OF D AND OF THE
C      VECTOR P ARE PRINTED AT EACH ITERATION.
C
C      DIMENSION X(M,L),    P(M)   ,W(M),    S(N,L),   E(N)
       DIMENSION X(100,12),P(100),W(100),S(10,12),E(10)
C      DIMENSION Q(N), T(L), SA(L), SB(L), G(M-N+1),Y(M-N+1,L)
       DIMENSION Q(10),T(12),SA(12),SB(12),G(100),   Y(100,12)
       LOGICAL ITRUE
       INTEGER P,Q,H,U,V,R
       KO=6
       KTMAX=100
       EPS=1.E-3
       DO 1 J=1,N
            Q(J)=0
     1 CONTINUE
       DO 2 I=1,M
            R=P(I)
            IF(R.LT.1.OR.R.GT.N) RETURN
            Q(R)=Q(R)+1
     2 CONTINUE
       DO 3 J=1,N
            IF(Q(J).EQ.0) RETURN
     3 CONTINUE
       ITRUE=.TRUE.
       I=0
       IT=0
       D=0.
       ITMAX=KTMAX
     4 DO 13 J=1,N
            IF(ITRUE) GOTO 5
            IF(J.NE.R) P(I)=J
            IF(J.EQ.R) P(I)=0
     5      V=0
            DO 7 H=1,M
                 IF(P(H).NE.J) GOTO 7
                 V=V+1
                 G(V)=W(H)
                 DO 6 K=1,L
                      Y(V,K)=X(H,K)
     6           CONTINUE
     7      CONTINUE
            IS=0
            CALL STANDN (V,L,Y,G,T,EPS,ITMAX,IS,F)
            ITMAX=KTMAX
            IF(.NOT.ITRUE) GOTO 9
            E(J)=F
            D=D+F
            DO 8 K=1,L
                 S(J,K)=T(K)
```

```
  8         CONTINUE
            GOTO 13
  9         IF(J.NE.R) GOTO 11
            A=F
            DO 10 K=1,L
                     SA(K)=T(K)
 10         CONTINUE
            GOTO 13
 11         IF(F.GT.B) GOTO 13
            B=F
            U=J
            DO 12 K=1,L
                     SB(K)=T(K)
 12         CONTINUE
 13 CONTINUE
    IF(.NOT.ITRUE) GOTO 14
    ITRUE=.FALSE.
    GOTO 17
 14 BU=B-E(U)
    RA=E(R)-A
    IF(BU.LT.RA) GOTO 15
    IT=IT+1
    P(I)=R
    GOTO 17
 15 IT=0
    E(R)=A
    E(U)=B
    D=D-RA+BU
    P(I)=U
    Q(R)=Q(R)-1
    Q(U)=Q(U)+1
    DO 16 K=1,L
            S(R,K)=SA(K)
            S(U,K)=SB(K)
 16 CONTINUE
    IF(IDR.EQ.1) WRITE(KO,18) I,D,(P(U),U=1,M)
 17 I=I+1
    IF(I.GT.M) I=I-M
    IF(IT.EQ.M) RETURN
    R=P(I)
    IF(Q(R).EQ.1) GOTO 17
    B=1.E30
 18 FORMAT(1X,I4,F12.2,23I3/(17X,23I3))
    GOTO 4
    END
```

Fig. U 13

The results in Fig. E9.1 are again for the example in Fig. B6, previously used in Figs. E5.1 and E6.

```
C
C     LOCATION-CLUSTERING WITH A GIVEN INITIAL PARTITION
C
      DIMENSION X(100,12),P(100),W(100),S(10,12),E(10)
      INTEGER P
      KI=5
      KO=6
    1 READ(KI,2) M,L,N1,N2,IDR,IP,IW
    2 FORMAT(16I5)
      IF(M.LE.1.OR.M.GT.100.OR.L.LT.1.OR.L.GT.12.OR.
     *   N1.GT.M.OR.N1.LT.1.OR.N2.LT.N1.OR.N2.GT.10) STOP
      WRITE(KO,3) M,L,N1,N2,IDR,IP,IW
    3 FORMAT('1','  M=',I3,'  L=',I2,'  N1=',I2,'  N2=',
     *       I2,'  IDR=',I1,'  IP=',I1,'  IW=',I1)
      WRITE(KO,4)
    4 FORMAT('0')
      DO 5 K=1,L
            READ(KI,6)  (X(I,K),I=1,M)
            WRITE(KO,7) (X(I,K),I=1,M)
    5 CONTINUE
    6 FORMAT(16F5.0)
    7 FORMAT(1X,10F7.1)
      IF(IW.NE.0) GOTO 9
      DO 8 I=1,M
            W(I)=1.
    8 CONTINUE
      GOTO 10
    9 READ(KI,6) (W(I),I=1,M)
      WRITE(KO,4)
      WRITE(KO,7)(W(I),I=1,M)
   10 DO 16 N=N1,N2
            WRITE(KO,4)
            IF(IP.NE.0) GOTO 12
            K=0
            DO 11 I=1,M
                  K=K+1
                  IF(K.GT.N) K=K-N
                  P(I)=K
   11       CONTINUE
            GOTO 13
   12       READ(KI,2)   (P(I),I=1,M)
   13       WRITE(KO,14) (P(I),I=1,M)
   14       FORMAT(1X,10I7)
            CALL CLUSTA (M,L,X,W,P,N,S,E,D,IDR)
            WRITE(KO,4)
            WRITE(KO,14) (P(I),I=1,M)
            WRITE(KO,4)
            DO 15 K=1,L
                  WRITE(KO,7) (S(J,K),J=1,N)
   15       CONTINUE
            WRITE(KO,4)
            WRITE(KO,7) (E(J),J=1,N),D
   16 CONTINUE
      GOTO 1
      END
```

Fig. H 9

```
M= 10  L= 2  N1= 2  N2= 5  IDR=0  IP=0  IW=0

  0.0     1.0     2.0     4.0     5.0     5.0     6.0     8.0     9.0    10.0
  1.0     2.0     0.0     8.0     7.0     9.0     7.0     4.0     3.0     5.0

    1       2       1       2       1       2       1       2       1       2

    1       1       1       2       2       2       2       2       2       2

  0.8     6.0
  1.2     7.0

  3.3    18.6    21.9

    1       2       3       1       2       3       1       2       3       1

    2       2       2       1       1       1       1       3       3       3

  5.0     0.8     8.8
  7.5     1.2     3.8

  4.2     3.3     3.3    10.9

    1       2       3       4       1       2       3       4       1       2

    3       3       3       4       4       4       4       1       1       2

  8.5    10.0     0.8     5.0
  3.5     5.0     1.2     7.5

  1.4     0.0     3.3     4.2     9.0

    1       2       3       4       5       1       2       3       4       5

    2       2       3       1       1       1       1       4       4       5

  5.0     0.5     2.0     8.5    10.0
  7.5     1.5     0.0     3.5     5.0

  4.2     1.4     0.0     1.4     0.0     7.1
```

Fig. E 9.1

The second example uses the coordinates of the 22 towns from Fig. T1. Fig. B23 contains the final allocations produced by KMEANS from the initial partition (3.2.10), and B24 those produced by CLUSTA. The centroids and the location centres, respectively, are marked by circles.

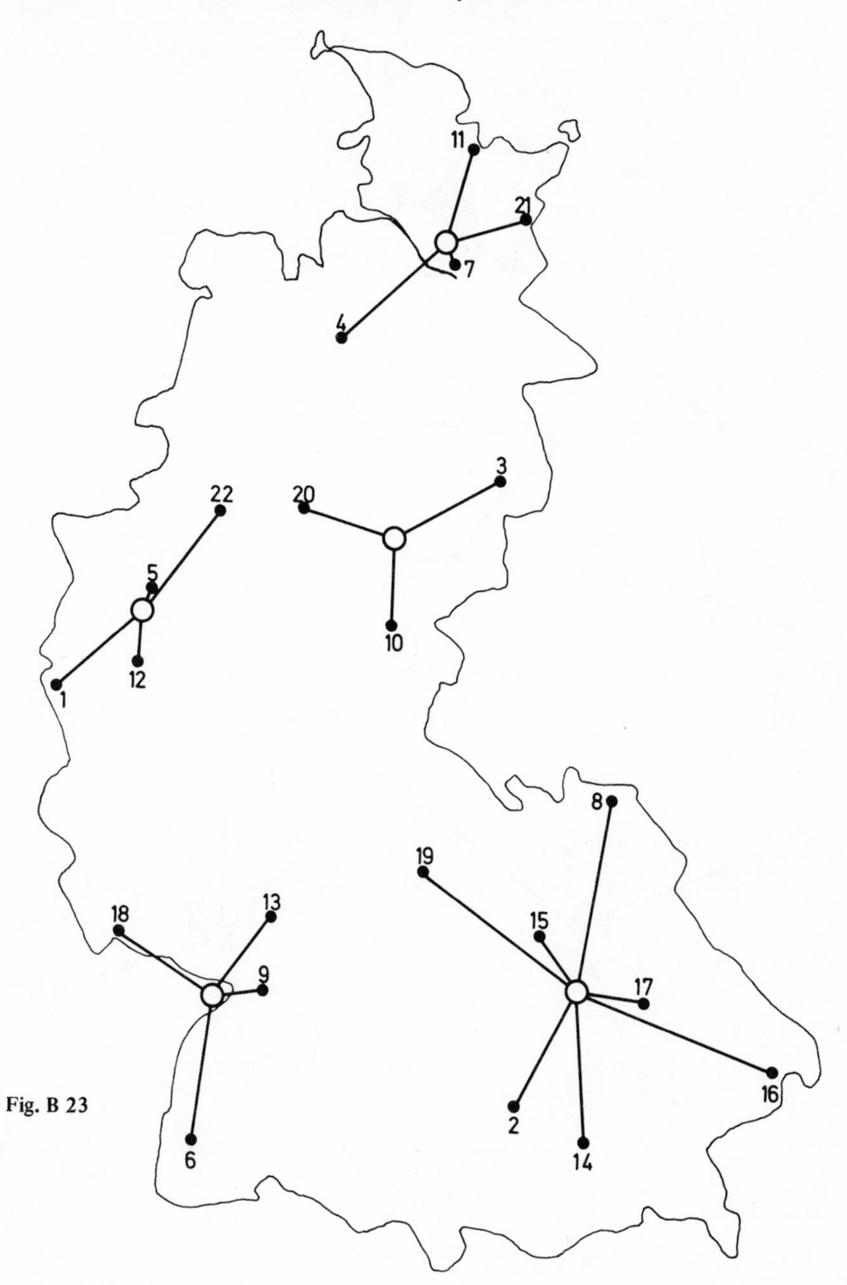

Fig. B 23

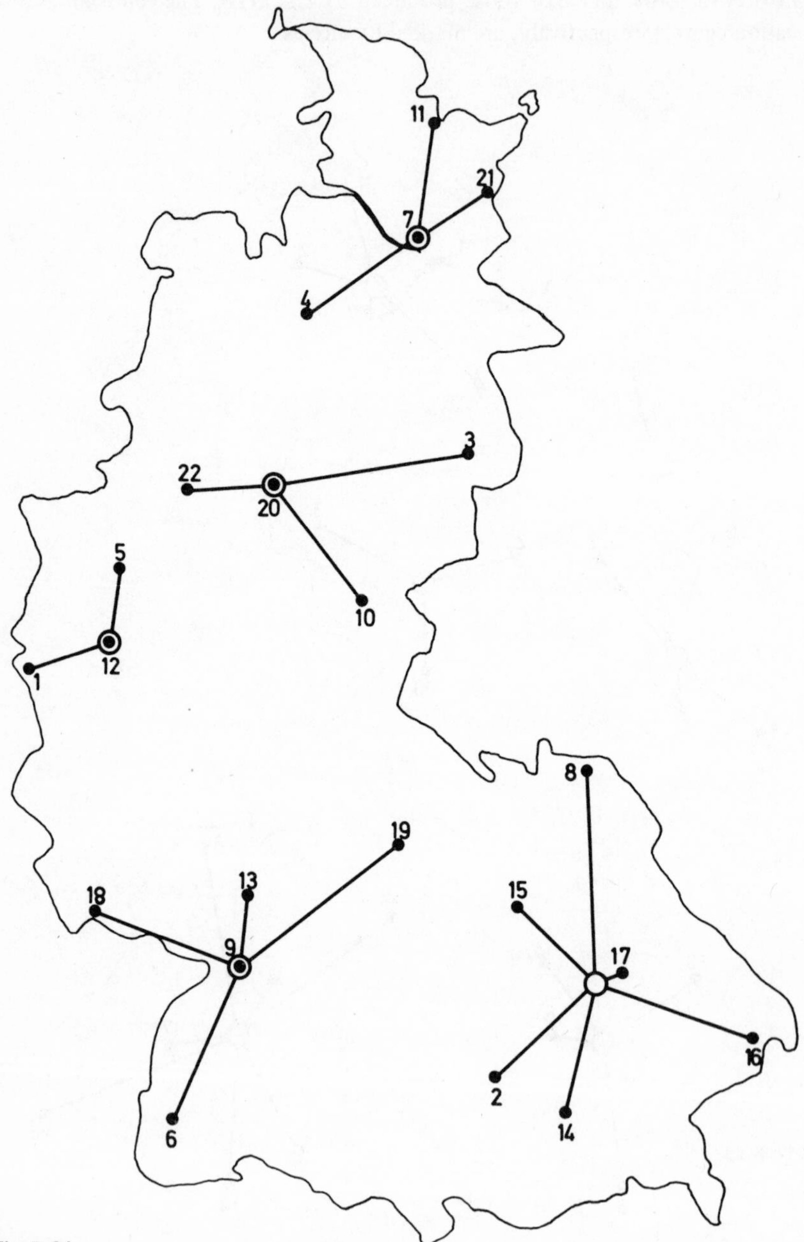

Fig. B 24

In the third example the coordinates of the 59 towns from E6.2 (Fig. B4) were used. The initial partition (3.2.10) was used on one occasion and on the other the final partition that resulted from this initial partition using KMEANS (KMEANS-start) was employed as the initial allocation for CLUSTA. Table T12 contains the related values of $\tilde{Z}_1$ for $n = 4, \ldots, 10$ for the final partitions obtained and, in the second case, also those location numbers which had still been assigned to other clusters.

Table T12

n	$\tilde{Z}_1$(3.2.10)	$\tilde{Z}_1$(KMEANS-Start)	Location number	
4	1 766	1 510	56	
5	1 317	1 323	56	
6	1 218	1 178	26	25
7	1 096	1 079	22	26
8	1 042	1 012	59	22
9	966	960	7	
10	907	905	15	

Usually KMEANS-start, which for $n = 1$ is identical to (3.4.5), provides the better values for the objective function $\tilde{Z}_1$. The process in this respect saves time since only very few of the cumbersome centre determinations using STANDN are necessary, while, with the standard initial partition (3.2.10), 6 to 7 passes were required, exactly as in both KMEANS and CLUSTA, and a case similar to T3 (left) would require over 60 iterative centre determinations instead of a maximum of two as in T12 (right).

Figs. B25 and B26 show the results for the 59 towns of the two initial value procedures for $n = 5$. The results of KMEANS-start for $n = 7$ are shown in Fig. B27 (as are also the numbers of the towns) and those for $n = 9$ are shown in Fig. B28 (see also Späth 1977b).

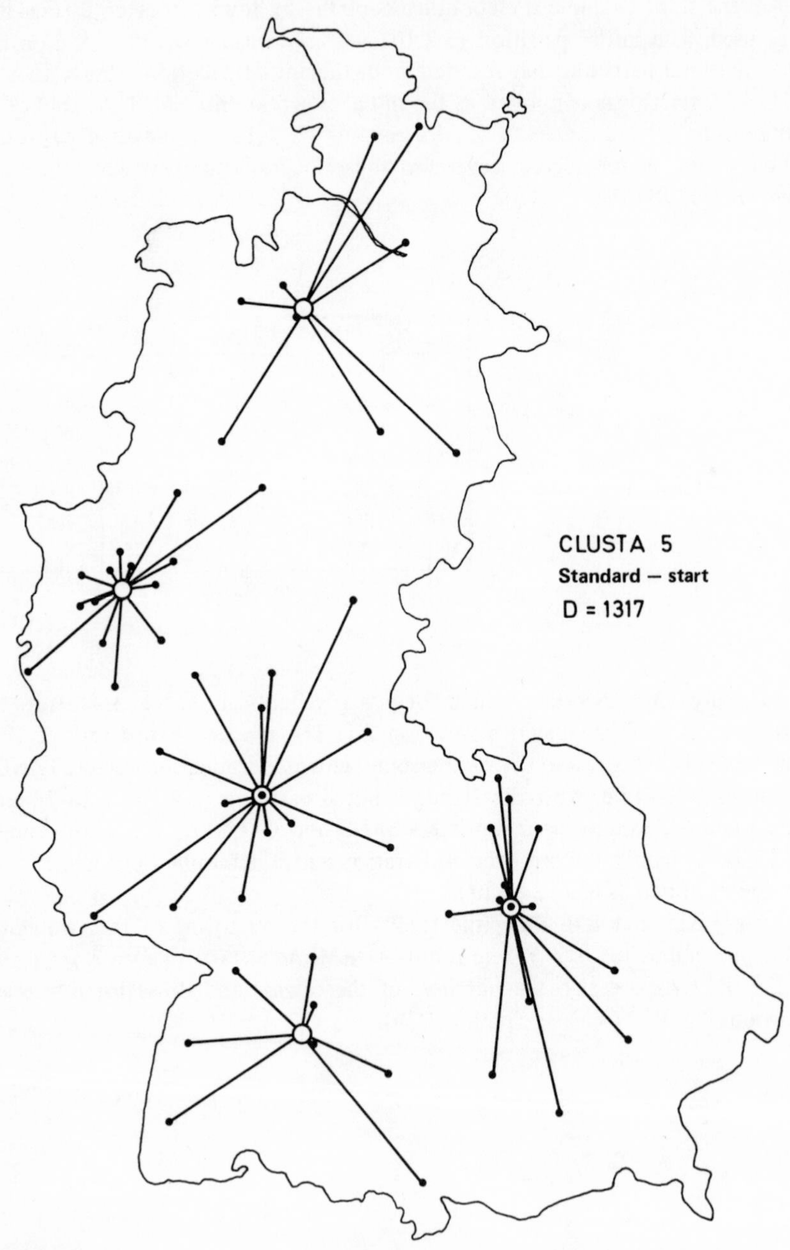
CLUSTA 5
Standard – start
D = 1317

Fig. B 25

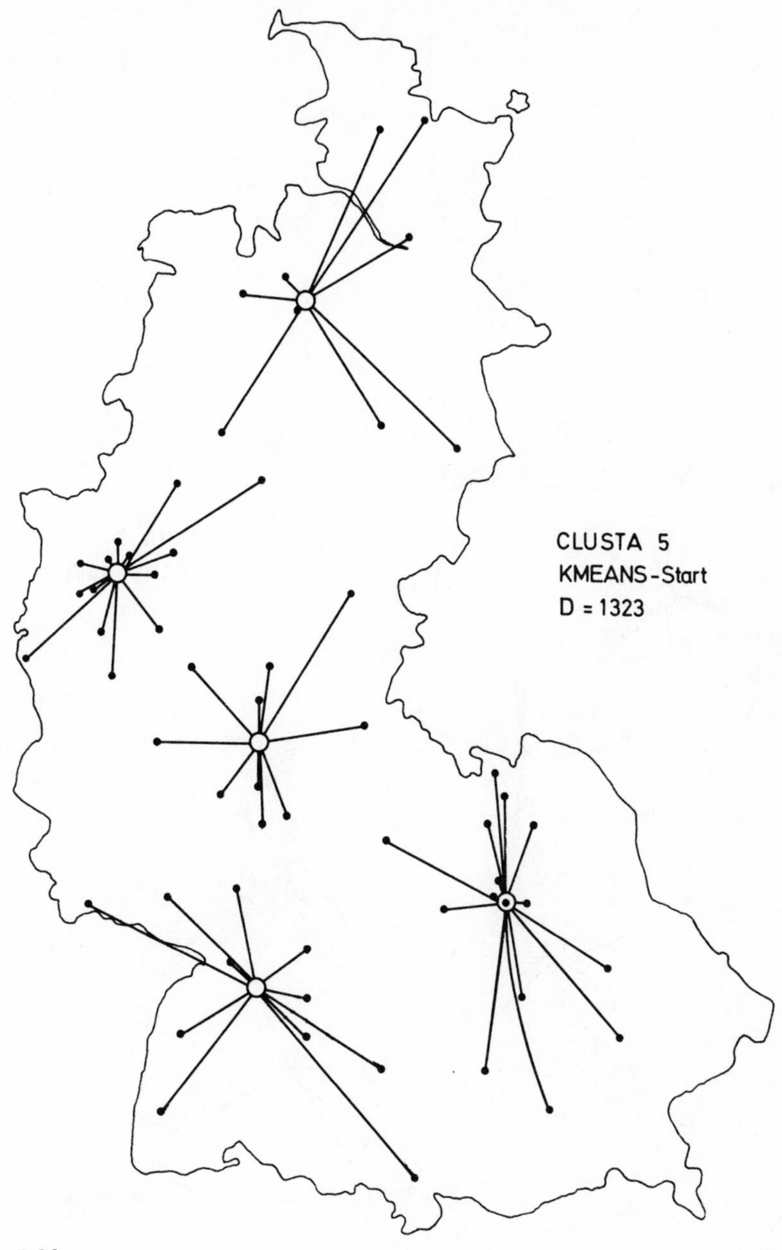

Fig. B 26

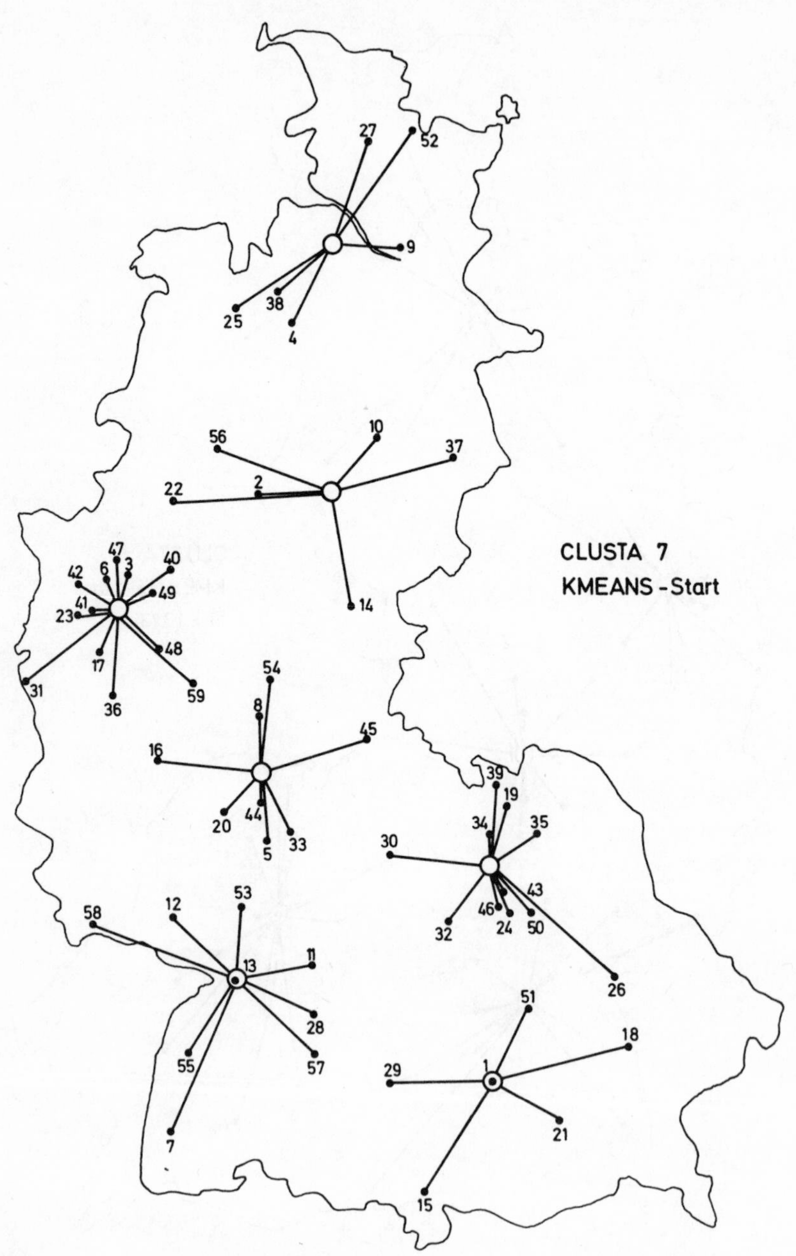

Fig. B 27

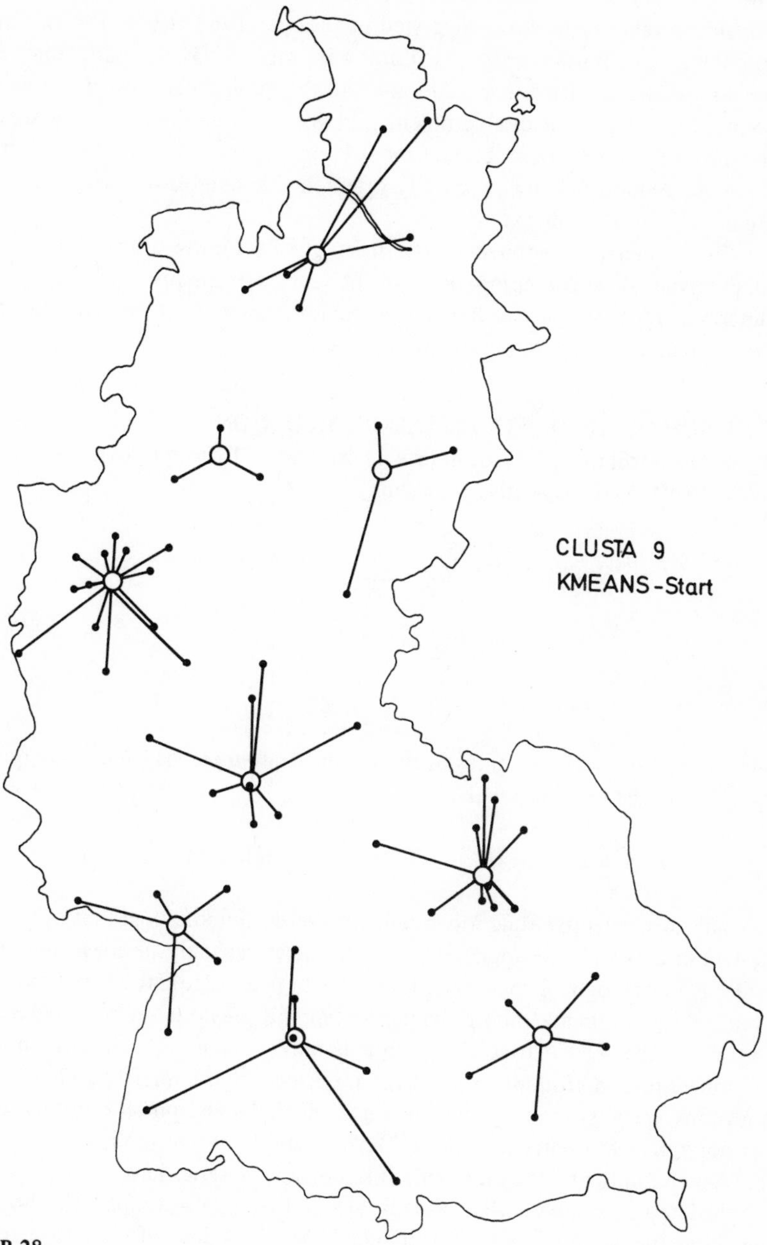

Fig. B 28

We believe that the KMEANS-start procedure is more effective than the various initial value procedures suggested in (Cooper 1963, 1964, 1967), which either determine initial optimal allocations on the assumption that the centres are identical with some locations, and then undertake the calculation of the centres for the inidividual problems, or work in a heuristic fashion similar to the procedures discussed in section 3.1. In (Cooper 1972) the objective function $\widetilde{Z}_1$ is modified by practically oriented side conditions, which are not of any interest in this context.

If the L_1 metric is applied to the right of (3.4.1), the solution for $n = 1$ becomes trivial, since the components of the vectors of the optimal location can be determined by simple median determination, and hence by sorting. The case $n > 1$ will be considered in the next section.

3.5 THE SUM OF ABSOLUTE DISTANCE CRITERION

Changing the norm in (3.2.5), or in (3.4.1) using $w_i = 1$, from L_2 to L_1, we obtain as in (Späth 1976) the objective function

$$Z_1^{(1)}(C_1, \ldots, C_n) = \sum_{j=1}^{n} \sum_{i \in C_j} \|x_i - s_j\|_1 = \sum_{j=1}^{n} \sum_{i \in C_j} \sum_{k=1}^{l} |x_{ik} - s_{jk}| \ . \tag{3.5.1}$$

Herein the s_{jk} are the coordinates of the within group medians which may not be unique but consist of an interval

$$s_{jk} = \left[u_{jk}, v_{jk}\right] \quad (j=1, \ldots, n, \ k=1, \ldots, l). \tag{3.5.2}$$

Minimising this objective function seems to be more suitable than that of (3.2.5) or (3.4.1) in a number of applications. If distances can only be measured rectilinearly, as in orthogonal road nets, then (3.5.1) is an adequate formulation for the multiple location-allocation problem from the previous section (see Späth 1976). When the data matrix (1.1) contains only ordinal data, then the means are no longer meaningful but the medians are. Especially for questionnaires where m questions are answered by persons on a bipolar scale, one has ordinal data values normally between 1 and 7 ($l = 7$). An example of this application is given in (Belschner and Späth 1977). Finally if seasonal time series have to be clustered for typical representatives, they usually are at first made comparable through giving them the same L_1 norm i.e. standardising the values of each time series such that their sum is 100%. Then here as in regression analysis (see Späth 1974) the L_1 metric is better used than the Euclidean one if outliers are likely to be in the curves: the corresponding large differences from the sought typical curves are not squared but come in only with their absolute value, thus reducing their influence. An example is given in (Späth 1978).

As a heuristic method for trying to minimise (3.5.1), one can use again the exchange method, as in KMEANS. The medians are determined for each component by sorting, and the cluster sums of absolute deviations from the medians

$$e_j = \sum_{i \in C_j} \|x_i - s_j\|_1 \tag{3.5.3}$$

are calculated for an initial partition such as (3.2.10). The details for updating the s_j and e_j values during the exchange process of objects are given in (Späth 1976). The corresponding subroutine EMEANS (Fig. U14) has nearly the same parameters as KMEANS. Only the former mean vectors S(J,*) have been replaced by the interval vector (U(J,*),V(J,*)) corresponding to (3.5.2), E(J) now corresponds to (3.5.3), and D means the value for the objective function (3.5.1) for the final partition.

Fig. U 14

```
      SUBROUTINE EMEANS (M,L,X,P,N,U,V,E,D,IDR)
C     DIMENSION X(M,L), P(M), U(N,L), V(N,L), E(N)
      DIMENSION X(100,12), P(100), U(10,12), V(10,12), E(10)
C     DIMENSION XX(M,L), Q(N), QQ(N+1), EVEN(N), Y(M-N)
      DIMENSION XX(100,12), Q(10), QQ(11), EVEN(10), Y(90)
C     DIMENSION UA(L), VA(L), UB(L), VB(L)
      DIMENSION UA(12), VA(12), UB(12), VB(12)
C     DIMENSION UF(L), VF(L), SA(L), SB(L), SF(L)
      DIMENSION UF(12), VF(12), SA(12), SB(12), SF(12)

      INTEGER Q,QQ,QJ,QQJ,QR,QQR,Q1,Q2,QW,T,
     *        P,Z,Z1,ZQ,W,R,G,SFK,SA,SB,SF
      LOGICAL EVEN,EVENJ,EVENR,ENDE,REVERS
C
C         CALCULATE NUMBER OF OBJECTS Q(J) FOR EACH INITIAL CLUSTER
C
      KO=6
      DO 1 J=1,N
          Q(J)=0
          E(J)=0.
    1 CONTINUE
      DO 2 I=1,M
          J=P(I)
          IF(J.LE.0.OR.J.GT.N) GOTO 37
          Q(J)=Q(J)+1
    2 CONTINUE
C
C         SORT FOR EACH J=1,N AND EACH K=1,L THE VALUES X(I,K)
C         WITH P(I)=J INTO Y(Z) WITH Z=Q(J) AND TRANSPORT IT
C         SEQUENTIALLY INTO THE ARRAY XX. FURTHER CALCULATE
C         THE VALUES E(J) AND D.
C
```

```
              D=0.
              QQ(1)=1
              DO 9 J=1,N
                   QJ=Q(J)
                   IF(QJ.EQ.0) GOTO 37
                   QQJ=QQ(J)
                   QQ(J+1)=QQJ+QJ
                   Q1=QJ/2
                   Q2=(QJ+1)/2
                   EVENJ=.FALSE.
                   IF(2*Q1.EQ.QJ) EVENJ=.TRUE.
                   EVEN(J)=EVENJ
              DO 8 K=1,L
                   Z=0
                   DO 3 I=1,M
                        IF(P(I).NE.J) GOTO 3
                        Z=Z+1
                        Y(Z)=X(I,K)
    3              CONTINUE
                   IF(QJ.EQ.1) GOTO 6
                   DO 5 W=2,QJ
                        ENDE=.TRUE.
                        QW=QJ-W+1
                        DO 4 Z=1,QW
                             Z1=Z+1
                             IF(Y(Z).LE.Y(Z1)) GOTO 4
                             H=Y(Z1)
                             Y(Z1)=Y(Z)
                             Y(Z)=H
                             ENDE=.FALSE.
    4                   CONTINUE
                        IF(ENDE) GOTO 6
    5              CONTINUE
    6              HH=Y(Q2)
                   U(J,K)=HH
                   V(J,K)=HH
                   IF(EVENJ) V(J,K)=Y(Q2+1)
                   H=0.
                   DO 7 Z=1,QJ
                        YZ=Y(Z)
                        ZQ=QQJ+Z-1
                        XX(ZQ,K)=YZ
                        H=H+ABS(YZ-HH)
    7              CONTINUE
                   E(J)=E(J)+H
                   D=D+H
    8         CONTINUE
    9 CONTINUE
      IF(N.EQ.1) GOTO 37
C
C        START OF EXCHANGE ALGORITHM
C
C        DELETE OBJECT I FROM CLUSTER R=P(I) IF Q(R)>1.
C
      I=0
      IT=0
   10 I=I+1
      IF(I.GT.M) I=I-M
      IF(IT.EQ.M) GOTO 37
      R=P(I)
      QR=Q(R)
      IF(QR.EQ.1) GOTO 10
```

```
      QQR=QQ(R)
      A=0.
      G=(QR+1)/2+QQR-1
      EVENR=EVEN(R)
      DO 17 K=1,L
             UR=U(R,K)
             VR=V(R,K)
             H= X(I,K)
             DO 11 W=1,QR
                    T=QR-W+QQR
                    IF(H.NE.XX(T.K)) GOTO 11
                    SA(K)=T
                    GOTO 12
 11          CONTINUE
 12          IF(EVENR) GOTO 15
             SS=XX(G-1,K)
             VV=XX(G+1,K)
             IF(H.NE.UR) GOTO 13
             UA(K)=SS
             VA(K)=VV
             GOTO 17
 13          IF(H.GT.UR) GOTO 14
             UA(K)=UR
             VA(K)=VV
             A=A+ABS(H-UR)
             GOTO 17
 14          UA(K)=SS
             VA(K)=UR
             A=A+ABS(H-UR)
             GOTO 17
 15          IF(H.GT.UR) GOTO 16
             UA(K)=VR
             VA(K)=VR
             A=A+ABS(H-VR)
             GOTO 17
 16          UA(K)=UR
             VA(K)=UR
             A=A+ABS(H-UR)
 17 CONTINUE
C
C        INSERT OBJECT I INTO THE CLUSTERS J=1,N EXCEPT J=R AND
C        DETERMINE THAT CLUSTER Z THAT GIVES THE LEAST INCREASE
C
      B=1.E30
      DO 26 J=1,N
             IF(J.EQ.R) GOTO 26
             F=0.
             QJ=Q(J)
             ENDE=QJ.GT.1
             QQJ=QQ(J)
             G=(QJ+1)/2+QQJ-1
             ZQ=QQ(J+1)
             EVENJ=EVEN(J)
             DO 24 K=1,L
                    UJ=U(J,K)
                    VJ=V(J,K)
                    H =X(I,K)
                    SFK=ZQ
                    DO 18 W=1,QJ
                           T=QQJ+W-1
                           IF(H.GT.XX(T,K))GOTO 18
                           SFK=T
                           GOTO 19
```

```
  18                CONTINUE
  19                SF(K)=SFK
                    IF(EVENJ) GOTO 21
                    SS=H
                    IF(H.GT.UJ) GOTO 20
                    UF(K)=H
                    IF(ENDE) SS=XX(G-1,K)
                    IF(H.LT.SS) UF(K)=SS
                    VF(K)=VJ
                    F=F+ABS(H-UJ)
                    GOTO 24
  20                IF(ENDE) SS=XX(G+1,K)
                    UF(K)=UJ
                    VF(K)=H
                    IF(H.GT.SS) VF(K)=SS
                    F=F+ABS(H-UJ)
                    GOTO 24
  21                IF(H.GE.VJ) GOTO 23
                    IF(H.LE.UJ) GOTO 22
                    UF(K)=H
                    VF(K)=H
                    GOTO 24
  22                UF(K)=UJ
                    VF(K)=UJ
                    F=F+ABS(H-UJ)
                    GOTO 24
  23                UF(K)=VJ
                    VF(K)=VJ
                    F=F+ABS(H-VJ)
  24          CONTINUE
              IF(F.GT.B) GOTO 26
              B=F
              Z=J
              DO 25 K=1,L
                    SB(K)=SF(K)
                    UB(K)=UF(K)
                    VB(K)=VF(K)
  25          CONTINUE
  26 CONTINUE
C
C       IF THE REDUCTION OF CLUSTER R IS GREATER THAN THE INCREASE
C       IN CLUSTER Z, THEN SHIFT OBJECT I AND UPDATE ALL
C       VALUES DEPENDING ON R AND Z.
C
     IF(B.LT.A) GOTO 27
     IT=IT+1
     GOTO 10
  27 E(R)=E(R)-A
     E(Z)=E(Z)+B
     D=D-A+B
     P(I)=Z
     Q(R)=Q(R)-1
     Q(Z)=Q(Z)+1
     EVEN(R)=.NOT.EVEN(R)
     EVEN(Z)=.NOT.EVEN(Z)
     REVERS=R.GT.Z
     DO 31 K=1,L
              U(R,K)=UA(K)
              V(R,K)=VA(K)
              U(Z,K)=UB(K)
              V(Z,K)=VB(K)
              JR=SA(K)
              JZ=SB(K)
```

```
            IF(REVERS) GOTO 29
            JZ1=JZ-1
            IF(JR.EQ.JZ1) GOTO 31
            JZ2=JZ1-1
            DO 28 W=JR,JZ2
                  XX(W,K)=XX(W+1,K)
28          CONTINUE
            XX(JZ1,K)=X(I,K)
            GOTO 31
29          JZ2=JZ+1
            H=XX(JZ,K)
            XX(JZ,K)=X(I,K)
            DO 30 W=JZ2,JR
                  HH=XX(W,K)
                  XX(W,K)=H
                  H=HH
30          CONTINUE
31 CONTINUE
   IF(REVERS) GOTO 33
   G=R+1
   DO 32 W=G,Z
         QQ(W)=QQ(W)-1
32 CONTINUE
   GOTO 35
33 G=Z+1
   DO 34 W=G,R
         QQ(W)=QQ(W)+1
34 CONTINUE
35 IF(IDR.NE.0) WRITE(KO,36) I,R,Z,A,B,E(R),E(Z),D
36 FORMAT(1X,3I4,5F13.4)
   GOTO 10
37 RETURN
   END
```

Fig. U 14

EMEANS can be used exactly in the same way as KMEANS: one can start with the initial partition (3.2.10), but it is recommended to use several ones at random and then to select that final partition with the minimal D value. For some applications it could be seen (see Späth 1977b, 1978) that it was very effective to take some final partition from KMEANS as the initial partition for EMEANS, just as for CLUSTA in the previous section.

A disadvantage of EMEANS as opposed to KMEANS is that twice as much storage is required for the data matrix (arrays X and XX). Remember that CLUDIA can be used as a substitute for EMEANS for smaller numbers of objects using a precalculated L_1 distance matrix. Of course, in this case the two subroutines EMEANS and CLUDIA deliver identical results – CLUDIA without the medians – for the same initial partition.

CHAPTER 4

Hierarchical Cluster Algorithms

4.1 DIVISIVE PROCEDURES

By the term 'hierarchical cluster algorithms' are understood (see Bock 1970) methods which for m given objects produce a sequence of partitions of lengths $n_1, \ldots, n_k$ such that

$$1 = n_0 < n_1 < n_2 < \ldots < n_k \leqq m \tag{4.1.1}$$

Thus sets of indices

$$\text{divisive} \downarrow \quad \begin{array}{l} C_1 = \{1, \ldots, m\} \\ C_1, C_2, \ldots, C_{n_1} \\ C_1, C_2, \ldots, C_{n_2} \\ \vdots \\ C_1, C_2, \ldots, C_{n_k} \end{array} \quad \uparrow \text{agglomerative} \tag{4.1.2}$$

are obtained wherein, with divisive procedures, clusters at level $j+1$ ($j=0, \ldots, k-1$) are produced by splitting up one or more of the clusters of level j; and, with agglomerative procedures, starting from $n_k = m$, clusters at level $j-1$ ($j = k, k-1, \ldots, 1$) are produced by merging clusters at level j. In both directions criteria are applied which define the algorithm. The splitting or merging strategies are usually designed optimally at each stage, but not the partition at a given level j (see Freitag 1972, Gower 1967). This is a disadvantage, as compared with such criteria as KMEANS, WMEANS, EMEANS or CLUDIA, of all hierarchical procedures (see Hodson 1970). However, especially with agglomerative procedures, their significantly shorter computation time up to $m = 100$ frequently compensates for this.

We shall discuss two divisive methods in this section and seven agglomerative methods in the next section.

The first method (MacQueen 1967), sets $n_j = j + 1$ ($j = 0, \ldots, m-1$). Starting with cluster C_1, which is split into two parts by applying the objective function in (3.2.5) with $n = 2$, at each successive step j ($j = 1, \ldots, m-1$) the cluster j_c with the largest sum of squared distances e_{jc}, as defined in (3.2.12), is divided into two clusters. It might be possible to do this exactly at any given stage, since there would be only $S(m_{jc}, 2) = 2^{m_{jc}-1} - 1$ partitions to be examined, which is also true of the second method. Given the practical sizes of m, however, the total enumeration might be too tedious, especially at the first stage. For this reason, subroutine DISMEA (Fig. U15) uses KMEANS with $n = 2$ to perform the division, with an initial partition (3.2.10); it must be made clear that optimal partitions will not necessarily be found at every step.

Fig. U 15

```
      SUBROUTINE DISMEA (M,L,X,P,R,NMAX,E,D,IDR)
C
C     CONSTRUCTION OF HIERARCHICAL CLUSTERS,
C
C     THE VECTORS X(M,*) ARE INITIALLY PARTITIONED INTO TWO
C     CLUSTERS WITH STANDARD INITIAL VALUES, ACCORDING TO
C     THE KMEANS PRINCIPLE,
C
C     AT EACH STAGE THE CLUSTER WITH THE GREATEST VARIANCE E(J)
C     IS AGAIN PARTITIONED INTO TWO CLUSTERS, UNTIL THE MAXIMUM
C     NUMBER NMAX <= M OF CLUSTERS IS REACHED,
C
C     THE PROCESS IS TERMINATED IF THE CLUSTER WITH THE GREATEST
C     SUM OF SQUARES POSSESSES ONLY TWO ELEMENTS,   IN THIS CASE
C     NMAX IS REDUCED,
C
C     P(I) INDICATES THE NUMBER OF THE CLUSTER TO WHICH X(I,*)
C     BELONGS,
C
C     R GIVES THE NUMBERS OF THE CLUSTERS IN THE ORDER IN WHICH
C     THEY WERE PARTITIONED,
C
C     IF IDR IS NON-ZERO, THE CURRENT VALUES OF THE VECTOR P
C     AND OF THE SUM OF SUM OF SQUARES D ARE PRINTED AT EACH
C     PARTITIONING,
C
C     NOTE THAT KMEANS DOES NOT NECESSARILY PRODUCE A
C     PARTITIONING WHICH IS OPTIMAL WITH RESPECT TO
C     ITS OBJECTIVE,
C
C     DIMENSION X(M,L),    P(M),   E(NMAX),R(NMAX)
      DIMENSION X(100,12),P(100),E(100) ,R(100)
C     DIMENSION XC(M,L),     PC(M),          SC(N,L)
      DIMENSION XC(100,12),PC(100),EC(2),SC(10,12)
      INTEGER P,PC,R
      KO=6
      IDRC=0
      IF(NMAX.LT.2.OR.NMAX.GT.M) RETURN
      D=0.
      E(1)=0.
      DO 1 I=1,M
           P(I)=1
    1 CONTINUE
      JC=1
      NC=1
      R(1)=1
```

```
    2 NC=NC+1
      IF(NC.GT.NMAX) RETURN
      MC=0
      KC=1
      DO 4  I=1,M
            IF(P(I).NE.JC) GOTO 4
            MC=MC+1
            PC(MC)=KC
            DO 3 K=1,L
                  XC(MC,K)=X(I,K)
    3       CONTINUE
            KC=KC+1
            IF(KC.GT.2) KC=1
    4 CONTINUE
      IF(MC.LE.2) GOTO 10
      CALL KMEANS (MC,L,XC,PC,2,SC,EC,DC,IDRC)
      MC=0
      DO 5  I=1,M
            IF(P(I).NE.JC) GOTO 5
            MC=MC+1
            IF(PC(MC).EQ.2) P(I)=NC
    5 CONTINUE
      D=D-E(JC)+DC
      IF(IDR.EQ.0) GOTO 8
      WRITE(KO,6) NC,(P(I),I=1,M)
    6 FORMAT('0',I3,' : ',15I3/(7X,15I3))
      WRITE(KO,7) D
    7 FORMAT('+',50X,F15.2)
    8 E(JC)=EC(1)
      E(NC)=EC(2)
      EMAX=0.
      DO 9  J=1,NC
            EJ=E(J)
            IF(EJ.LE.EMAX) GOTO 9
            JC=J
            EMAX=EJ
    9 CONTINUE
      R(NC)=JC
      GOTO 2
   10 NMAX=NC-1
      END
```

Fig. U 15

In DISMEA the maximum number of clusters required, NMAX, can be specified, on condition that NMAX ⩽ M. The subroutine writes into the array R(NMAX) the serial numbers of the clusters which are split up, in the order in which this occurs. When a cluster with the greatest sum of squared distances is split up, the index of the next unfilled element of R becomes the next serial number to be used. The E(NMAX+1) array contains the corresponding sums of squares, and their sum. We shall now look at examples, calculated using main program H10; in this program, N1 has no meaning and is included only to maintain compatibility of input between several programs.

```
C
C     DIVISIVE HIERARCHICAL CLUSTERING BY THE USE OF KMEANS
C
      DIMENSION X(100,12),P(100),E(100),R(100)
      INTEGER P,R
      KI=5
      KO=6
    1 READ(KI,2) M,L,N1,NMAX,IDR
    2 FORMAT(16I5)
      IF(M.LE.1.OR.M.GT.100.OR.L.LT.1.OR.L.GT.12.OR.
     *    NMAX.LT.2.OR.NMAX.GT.M) STOP
      IDR=1
      WRITE(KO,3) M,L,NMAX,IDR
    3 FORMAT('1','  M=',I3,'  L=',I2,'  NMAX=',I2,
     *    '    IDR=',I1)
      WRITE(KO,4)
    4 FORMAT('0')
      DO 5 K=1,L
           READ(KI,6)   (X(I,K),I=1,M)
           WRITE(KO,7)  (X(I,K),I=1,M)
    5 CONTINUE
    6 FORMAT(16F5.0)
    7 FORMAT(1X,10F7.1)
      WRITE(KO,4)
      CALL DISMEA (M,L,X,P,R,NMAX,E,D,IDR)
      WRITE(KO,4)
      WRITE(KO,8) (P(I),I=1,M)
    8 FORMAT(7X,15I3)
      WRITE(KO,4)
      NM=NMAX-1
      WRITE(KO,8) (R(J),J=1,NM)
      WRITE(KO,4)
      WRITE(KO,9) (E(J),J=1,NMAX),D
    9 FORMAT(1X,5F13.2)
      GOTO 1
      END
```

Fig. H 10

The first example (Fig. E10.1) takes the 23 seasonal curves Fig. E6.3, using H10 with IDR = 1. The values $n_1 = 2, \ldots, n_8 = 9$, with colons, are printed on the left. Next comes the usual vector p_i $(i = 1, \ldots, m)$ containing the serial numbers of the clusters to which the objects are allocated. The D values corresponding to $n_j = j + 1$ are on the extreme right. The last vector P which is obtained is followed by the vector R, then the array E, and the final value of D. From the earlier explanation, the values in the vector R mean that the first cluster to be split is no. 1, then no. 1 again, then the cluster which has meanwhile been given the serial number 2, then no. 3, and so on. Throughout this process a new cluster is always allocated the next available serial number. All this will be made clear diagrammatically in the second example.

Fig. E 10.1

```
M= 23   L=12   NMAX= 9   IDR=1

   3.0     3.0     3.0     4.0     4.0     4.0     0.0     1.0     0.0     0.0
   4.0     5.0     4.0     4.0     4.0    21.0    19.0    19.0     8.0     9.0
  13.0    15.0     4.0
   9.0    12.0    13.0    11.0     9.0     8.0     2.0     2.0     2.0     3.0
   2.0     1.0     1.0     1.0     1.0    13.0    15.0    13.0     8.0     7.0
  12.0    13.0     4.0
  22.0    21.0    23.0    20.0    20.0    21.0     5.0     6.0     6.0     7.0
   2.0     2.0     1.0     2.0     1.0    10.0     8.0    14.0     9.0     9.0
  11.0    12.0     6.0
  14.0    16.0    15.0    15.0    14.0    15.0     9.0     8.0    10.0     9.0
   5.0     5.0     3.0     3.0     4.0     6.0     6.0     9.0     8.0     9.0
  10.0    10.0     7.0
  13.0    12.0    12.0    12.0    13.0    12.0    14.0    15.0    13.0    14.0
   8.0     7.0     6.0     6.0     8.0     2.0     2.0     7.0     8.0     8.0
   9.0     9.0     7.0
  11.0    10.0    10.0    11.0    11.0    10.0    20.0    19.0    21.0    19.0
   9.0     9.0     9.0    10.0    10.0     0.0     1.0     3.0     9.0     8.0
   9.0     8.0     8.0
  10.0    10.0     9.0     9.0     9.0     9.0    20.0    22.0    19.0    19.0
  10.0    11.0    10.0    10.0    11.0     1.0     0.0     0.0     8.0     7.0
   8.0     9.0     8.0
   8.0     6.0     6.0     8.0     7.0     8.0    14.0    12.0    15.0    13.0
  12.0    13.0    12.0    12.0    13.0     1.0     1.0     1.0     8.0     9.0
   7.0     8.0    10.0
   4.0     3.0     3.0     5.0     5.0     5.0     9.0    10.0     8.0     8.0
  15.0    14.0    15.0    16.0    14.0     4.0     4.0     3.0     9.0     9.0
   7.0     6.0    10.0
   1.0     2.0     1.0     0.0     2.0     2.0     5.0     4.0     4.0     4.0
  21.0    20.0    23.0    21.0    22.0     8.0    10.0     4.0     8.0     6.0
   6.0     5.0    11.0
   1.0     1.0     1.0     2.0     1.0     2.0     2.0     1.0     1.0     3.0
   8.0     9.0    13.0    12.0     9.0    15.0    12.0     8.0     8.0    10.0
   4.0     3.0    12.0
   4.0     4.0     4.0     3.0     5.0     4.0     0.0     0.0     1.0     1.0
   4.0     4.0     3.0     3.0     3.0    19.0    22.0    19.0     9.0     9.0
   4.0     2.0    13.0


2 :    2  2  2  2  2  2  2  2  2  2  1  1  1  1  1
       1  1  1  1  1  2  2  1                             5091.90

3 :    2  2  2  2  2  2  2  2  2  2  1  1  1  1  1
       3  3  3  3  3  2  2  1                             3036.90

4 :    4  4  4  4  4  4  2  2  2  2  1  1  1  1  1
       3  3  3  3  3  4  4  1                             1485.99

5 :    4  4  4  4  4  4  2  2  2  2  1  1  1  1  1
       5  5  5  3  3  4  4  1                              883.39

6 :    6  6  6  6  6  6  2  2  2  2  1  1  1  1  1
       5  5  5  3  3  4  4  1                              469.14

7 :    6  6  6  6  6  6  2  2  2  2  1  1  1  1  1
       5  5  5  3  3  4  4  7                              238.87

8 :    6  6  6  6  6  6  2  2  2  2  1  1  1  1  1
       8  8  5  3  3  4  4  7                              154.87

9 :    9  6  6  9  9  9  2  2  2  2  1  1  1  1  1
       8  8  5  3  3  4  4  7                              126.87
```

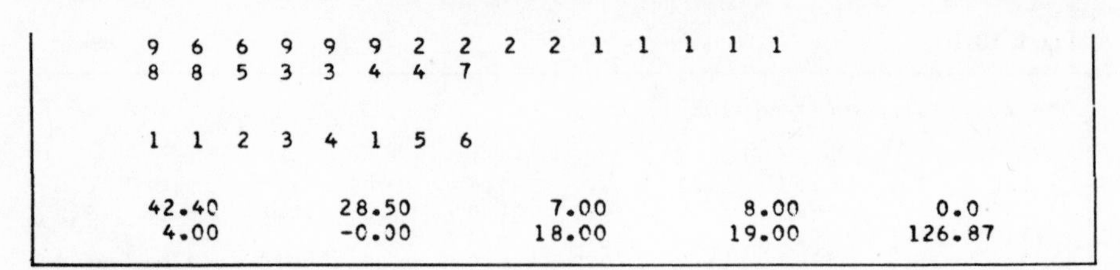

```
9  6  6  9  9  9  2  2  2  2  1  1  1  1  1
8  8  5  3  3  4  4  7

1  1  2  3  4  1  5  6

42.40          28.50          7.00          8.00          0.0
 4.00          -0.00         18.00         19.00        126.87
```

Fig. E 10.1

The second example E10.2 once more uses the cordinates of the 59 towns in Fig. B5. The output is interpreted in the same way as Fig. E10.1. The successive subdivisions of the clusters, corresponding to the vector R, are shown in Fig. B29. The cluster serial numbers in the vector refer to the lowest levels at each stage of the branching tree in Fig. B29. Fig. B30 shows the division of the towns into the ten clusters. The division can be reconstructed by using the vector R, as in Fig. B29.

Fig. E 10.2

```
M= 59   L= 2   NMAX=10   IDR=1

  54.0     0.0   -31.0     8.0     1.0   -36.0   -22.0     0.0    34.0    28.0
  12.0   -21.0    -6.0    21.0    38.0   -24.0   -38.0    86.0    58.0    -9.0
  70.0   -20.0   -43.0    59.0    -5.0    83.0    27.0    12.0    30.0    31.0
 -57.0    44.0     7.0    54.0    65.0   -35.0    46.0     5.0    56.0   -21.0
 -40.0   -43.0    57.0     0.0    25.0    56.0   -34.0   -24.0   -25.0    64.0
  63.0    37.0    -5.0     2.0   -18.0   -10.0    12.0   -40.0   -16.0
 -65.0    71.0    53.0   111.0    -9.0    52.0   -76.0    20.0   129.0    84.0
 -38.0   -26.0   -41.0    45.0   -90.0    10.0    35.0   -57.0    -1.0    -3.0
 -74.0    70.0    44.0   -26.0   114.0   -41.0   153.0   -49.0   -65.0   -12.0
  28.0   -28.0    -7.0    -8.0    -8.0    25.0    79.0   118.0     4.0    54.0
  45.0    51.0   -21.0     0.0    15.0   -25.0    56.0    36.0    49.0   -26.0
 -48.0   155.0   -24.0    28.0   -58.0    82.0   -58.0   -28.0    28.0

 2 :    1  2  2  2  1  2  1  2  2  2  1  1  1  2  1
        2  2  1  1  1  1  2  2  1  2  1  2  1  1  1
        2  1  1  1  1  2  2  2  1  2  2  2  1  1  1
        1  2  2  2  1  1  2  1  2  1  2  1  1  2          121423.88

 3 :    1  3  3  2  1  3  1  3  2  2  1  1  1  3  1
        3  3  1  1  1  1  3  3  1  2  1  2  1  1  1
        3  1  1  1  1  3  2  2  1  3  3  3  1  1  1
        1  3  3  3  1  1  2  1  3  1  3  1  1  3           78126.50

 4 :    1  3  3  2  4  3  4  3  2  2  4  4  4  3  1
        3  3  1  1  4  1  3  3  1  2  1  2  4  1  1
        3  1  4  1  1  3  2  2  1  3  3  3  1  4  4
        1  3  3  3  1  1  2  4  3  4  3  4  4  3           49600.51

 5 :    5  3  3  2  4  3  4  3  2  2  4  4  4  3  5
        3  3  5  1  4  5  3  3  1  2  5  2  4  5  1
        3  1  4  1  1  3  2  2  1  3  3  3  1  4  4
        1  3  3  3  1  5  2  4  3  4  3  4  4  3           40052.36

 6 :    5  3  3  2  4  3  4  6  2  2  4  4  4  6  5
        6  3  5  1  4  5  3  3  1  2  5  2  4  5  1
        3  1  4  1  1  6  2  2  1  3  3  3  1  4  4
        1  3  6  3  1  5  2  4  6  4  3  4  4  6           35282.91
```

```
 7 :   5  3  3  2  7  3  4  6  2  2  4  7  4  6  5
       6  3  5  1  7  5  3  3  1  2  5  2  4  5  1
       3  1  7  1  1  6  2  2  1  3  3  3  1  7  7
       1  3  6  3  1  5  2  7  6  4  3  4  7  6          28875.05

 8 :   5  3  3  2  7  3  4  6  8  2  4  7  4  6  5
       6  3  5  1  7  5  3  3  1  2  5  8  4  5  1
       3  1  7  1  1  6  2  2  1  3  3  3  1  7  7
       1  3  6  3  1  5  8  7  6  4  3  4  7  6          24671.10

 9 :   5  3  9  2  7  9  4  6  8  2  4  7  4  6  5
       6  9  5  1  7  5  3  9  1  2  5  8  4  5  1
       9  1  7  1  1  6  2  2  1  9  9  9  1  7  7
       1  9  6  9  1  5  8  7  6  4  3  4  7  6          21251.48

10 :  10  3  9  2  7  9  4  6  8  2  4  7  4  6 10
       6  9  5  1  7  5  3  9  1  2  5  8  4 10  1
       9  1  7  1  1  6  2  2  1  9  9  9  1  7  7
       1  9  6  9  1  5  8  7  6  4  3  4  7  6          18595.21


      10  3  9  2  7  9  4  6  8  2  4  7  4  6 10
       6  9  5  1  7  5  3  9  1  2  5  8  4 10  1
       9  1  7  1  1  6  2  2  1  9  9  9  1  7  7
       1  9  6  9  1  5  8  7  6  4  3  4  7  6


       1  2  1  1  3  4  2  3  5


    2117.22       2999.99       288.68       2222.68       963.00
    3000.49       4160.98       471.35       1655.70       715.33
   18595.21
```

Fig. E 10.2

R = (1 2 1 1 3 4 2 3 5)

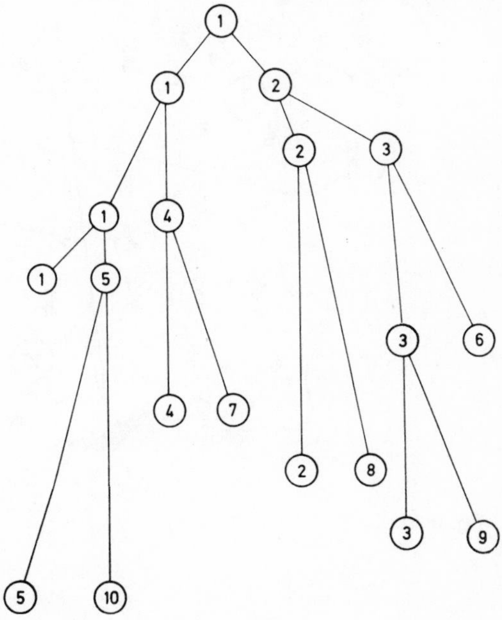

Fig. B 29

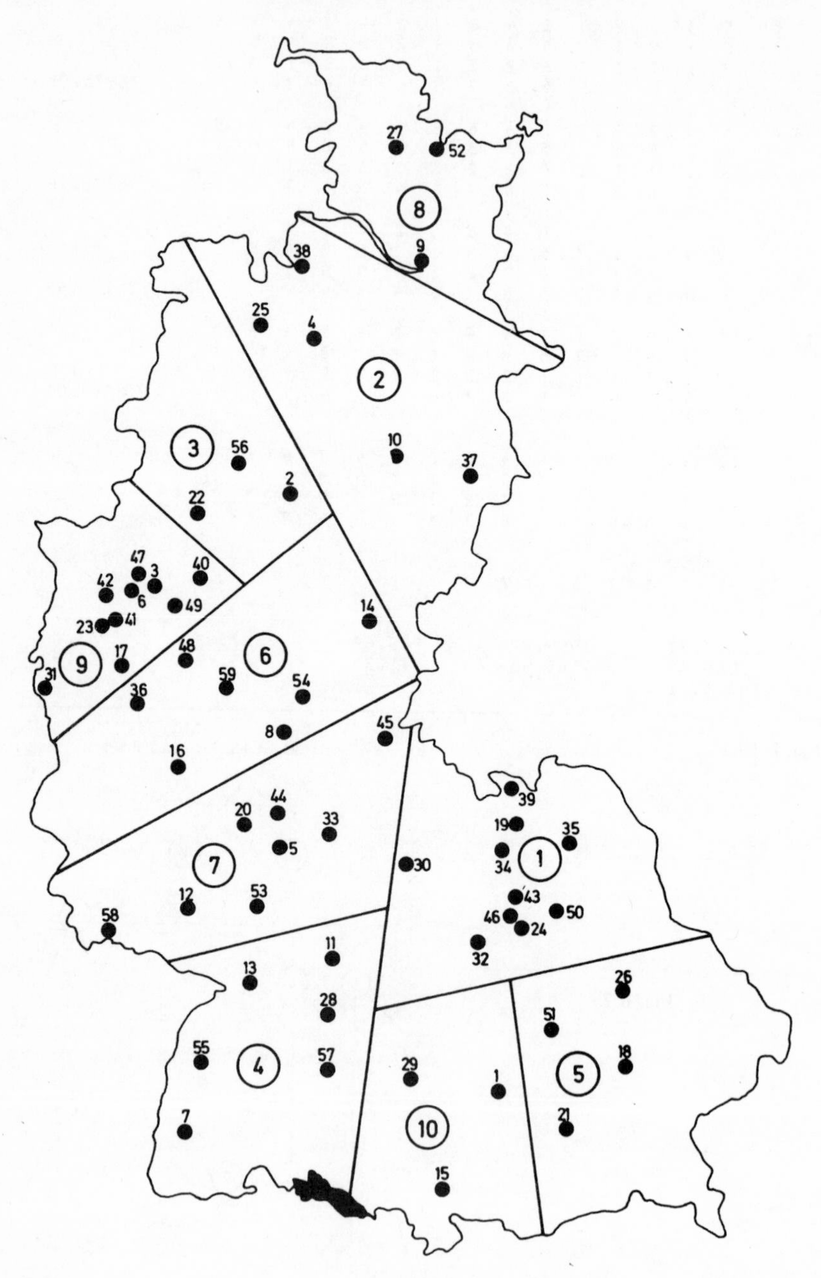
27
52
8
9
38
25
4
2
3
56
10
37
2
22
47
3
40
42
6
49
23
41
14
9
17
48
6
31
59
54
36
8
45
16
44
20
33
5
7
30
39
19
35
34
1
43
46
50
24
12
53
58
32
11
13
28
26
51
55
4
57
18
5
29
1
7
10
21
15

Fig. B 30

In the second method (Edwards and Cavalli-Sforza 1965) $n_j = 2^j$ ($j = 1, ..., \text{NMAX}$; $2^{\text{NMAX}} \leqslant m$). Here each cluster is divided up into two new clusters at each step, until a step results in the appearance of a one-element cluster, which cannot be subdivided.

One way to proceed is as in DISMEA, and, each time a cluster C_p is divided, to try to minimise the objective function

$$\sum_{j=1}^{2} \sum_{i \in C_j} \|x_i - \bar{x}_j\|^2 \quad (i \in C_p) \tag{4.1.3}$$

or – which is equivalent to (3.2.4) – maximise the objective function

$$\sum_{j=1}^{2} m_j \|\bar{x}_j - \bar{x}\|^2 \quad (i \in C_p) \tag{4.1.4}$$

By making use of $m\bar{x} = m_1 \bar{x}_1 + m_2 \bar{x}_2$, (4.1.4) can be converted into

$$\frac{m_1 m_2}{m_1 + m_2} \|\bar{x}_1 - \bar{x}_2\|^2 \tag{4.1.5}$$

so that a weighted squared distance between centroids would always be maximised while (4.1.3) was being minimised.

An alternative is to try to maximise the unweighted centroid distance

$$\|\bar{x}_1 - \bar{x}_2\| \tag{4.1.6}$$

at each division.

Instead of using either of these two principles, we shall use at each division the subroutine ZWEIGO (Fig. U4), which operates heuristically. Subroutine DIVGOW (Fig. U16) produces a maximum of NMAX clusters by subdivision. Dividing ceases if one of the clusters to be divided contains fewer than three elements. If all clusters are split into two parts for the jth time, they are numbered by $2^j - 1, \ldots, 2^{j+1} - 2$: if the divisions are no longer possible the numbering is transformed so that it begins with 1 – possibly already in the previous division level. The SD(NMAX–1) array contains the values (4.1.6) for the successive divisions.

Fig. U 16

```
      SUBROUTINE DIVGOW (M,L,X,P,NMAX,SD,IDR)
C
C     CONSTRUCTION OF HIERARCHICAL CLUSTERS,
C
C     THE VECTORS X(M,*) ARE INITIALLY DIVIDED INTO TWO CLUSTERS
C     BY USING ZWEIGO,
C
C     EACH CLUSTER IS SUCCESSIVELY RE-DIVIDED USING ZWEIGO, UNTIL
C     EITHER NMAX <= M CLUSTERS ARE REACHED (IN THIS SENSE NMAX
C     IS EFFECTIVELY A POWER OF 2) OR A CLUSTER WITH FEWER THAN 3
C     ELEMENTS IS ENCOUNTERED (IN WHICH CASE NMAX IS REDUCED),
C
C     P(I) CONTAINS THE NUMBER OF THE CLUSTER TO WHICH X(I,*)
C     BELONGS,   SD(NMAX-1) CONTAINS THE DISTANCES OF THE
C     CENTROID, AS SUCCESSIVELY ENCOUNTERED.
C
C     IF IDR IS NON-ZERO, THE CURRENT VALUES OF THE VECTOR P ARE
C     PRINTED AFTER EACH DIVISION,
C
C     DIMENSION X(M,L),    P(M)   ,SD(2*NMAX)
      DIMENSION X(100,12),P(100),SD(200)
C     DIMENSION X1(M,L)    ,X2(M,L),    P1(M)   ,P2(M)
      DIMENSION X1(100,12),X2(100,12),P1(100),P2(100)
C     DIMENSION S1(L) ,S2(L)
      DIMENSION S1(12),S2(12)
      INTEGER P,P1,P2,R
      KO=6
      IDRC=0
      IF(NMAX.LT.2.OR.NMAX.GT.M) RETURN
      CALL ZWEIGO (M,L,X,P,S1,S2,SD(1),IDRC)
      NC=2
      JC=2
      ND=1
    1 IF(IDR.EQ.0) GOTO 3
      WRITE(KO,2) NC,(P(I),I=1,M)
    2 FORMAT('0',I3,' : ',15I3/(7X,15I3))
    3 DO 11 J=2,NC,2
            J1=J+NC-3
            J2=J1+1
            M1=0
            M2=0
            DO 7 I=1,M
                  R=P(I)
                  IF(R.NE.J1) GOTO 5
                  M1=M1+1
                  DO 4 K=1,L
                        X1(M1,K)=X(I,K)
    4             CONTINUE
                  GOTO 7
    5             IF(R.NE.J2) GOTO 7
                  M2=M2+1
                  DO 6 K=1,L
                        X2(M2,K)=X(I,K)
    6             CONTINUE
```

```
  7         CONTINUE
            IF(M1.LE.2.OR.M2.LE.2) GOTO 12
            ND=ND+1
            CALL ZWEIGO (M1,L,X1,P1,S1,S2,SD(ND),IDRC)
            ND=ND+1
            CALL ZWEIGO (M2,L,X2,P2,S1,S2,SD(ND),IDRC)
            L1=1
            L2=1
            DO 10 I=1,M
                  R=P(I)
                  IF(R.NE.J1) GOTO 8
                  R=JC+P1(L1)
                  L1=L1+1
                  GOTO 9
  8               IF(R.NE.J2) GOTO 10
                  R=JC+2+P2(L2)
                  L2=L2+1
  9               P(I)=R
 10         CONTINUE
            JC=JC+4
 11   CONTINUE
      NC=NC+NC
      IF(NC.LT.NMAX) GOTO 1
 12   NMAX=ND+1
      NJ=ND-1
      DO 13 I=1,M
            P(I)=P(I)-NJ
 13   CONTINUE
      RETURN
      END
```

Fig. U 16

The first example, Fig. E11.1, computed with the help of main program H11, again uses the 23 seasonal curves. The results, when compared with Fig. E10.1, display considerable differences, arising on the one hand from the differing strategies, and on the other from the relatively large number of clusters compared with the number $m = 23$ of objects.

Fig. H 11

```
C
C     DIVISIVE HIERARCHICAL CLUSTERING BY THE USE OF ZWEIGO
C
      DIMENSION X(100,12),P(100),SD(200)
      INTEGER P
      KI=5
      KO=6
    1 READ(KI,2) M,L,N1,NMAX,IDR
    2 FORMAT(16I5)
      IDR=1
      IF(M.LE.2.OR.M.GT.100.OR.L.LT.1.OR.L.GT.12) STOP
      WRITE(KO,3) M,L,NMAX,IDR
    3 FORMAT('1',' M=',I3,' L=',I2,' NMAX=',I2,
     *      ' IDR=',I1)
      WRITE(KO,4)
```

```
    4 FORMAT('0')
      DO 5 K=1,L
           READ(KI,6)  (X(I,K),I=1,M)
           WRITE(KO,7) (X(I,K),I=1,M)
    5 CONTINUE
    6 FORMAT(16F5.0)
    7 FORMAT(1X,10F7.1)
      WRITE(KO,4)
      CALL DIVGOW (M,L,X,P,NMAX,SD,IDR)
      WRITE(KO,4)
      WRITE(KO,8) (P(I),I=1,M)
    8 FORMAT(7X,15I3)
      WRITE(KO,4)
      ND=NMAX-1
      WRITE(KO,9) (SD(J),J=1,ND)
    9 FORMAT(1X,5F13.2)
      GOTO 1
      END
```

Fig. H 11

Fig. E 11.1

```
M= 23   L=12   NMAX= 9   IDR=1

   3.0    3.0    3.0    4.0    4.0    4.0    0.0    1.0    0.0    0.0
   4.0    5.0    4.0    4.0    4.0   21.0   19.0   19.0    8.0    9.0
  13.0   15.0    4.0
   9.0   12.0   13.0   11.0    9.0    8.0    2.0    2.0    2.0    3.0
   2.0    1.0    1.0    1.0    1.0   13.0   15.0   13.0    8.0    7.0
  12.0   13.0    4.0
  22.0   21.0   23.0   20.0   20.0   21.0    5.0    6.0    6.0    7.0
   2.0    2.0    1.0    2.0    1.0   10.0    8.0   14.0    9.0    9.0
  11.0   12.0    6.0
  14.0   16.0   15.0   15.0   14.0   15.0    9.0    8.0   10.0    9.0
   5.0    5.0    3.0    3.0    4.0    6.0    6.0    9.0    8.0    9.0
  10.0   10.0    7.0
  13.0   12.0   12.0   12.0   13.0   12.0   14.0   15.0   13.0   14.0
   8.0    7.0    6.0    6.0    8.0    2.0    2.0    7.0    8.0    8.0
   9.0    9.0    7.0
  11.0   10.0   10.0   11.0   11.0   10.0   20.0   19.0   21.0   19.0
   9.0    9.0    9.0   10.0   10.0    0.0    1.0    3.0    9.0    8.0
   9.0    8.0    8.0
  10.0   10.0    9.0    9.0    9.0    9.0   20.0   22.0   19.0   19.0
  10.0   11.0   10.0   10.0   11.0    1.0    0.0    0.0    8.0    7.0
   8.0    9.0    8.0
   8.0    6.0    6.0    8.0    7.0    8.0   14.0   12.0   15.0   13.0
  12.0   13.0   12.0   12.0   13.0    1.0    1.0    1.0    8.0    9.0
   7.0    8.0   10.0
   4.0    3.0    3.0    5.0    5.0    5.0    9.0   10.0    8.0    8.0
  15.0   14.0   15.0   16.0   14.0    4.0    4.0    3.0    9.0    9.0
   7.0    6.0   10.0
   1.0    2.0    1.0    0.0    2.0    2.0    5.0    4.0    4.0    4.0
  21.0   20.0   23.0   21.0   22.0    8.0   10.0    4.0    8.0    6.0
   6.0    5.0   11.0
   1.0    1.0    1.0    2.0    1.0    2.0    2.0    1.0    1.0    3.0
   8.0    9.0   13.0   12.0    9.0   15.0   12.0    8.0    8.0   10.0
   4.0    3.0   12.0
   4.0    4.0    4.0    3.0    5.0    4.0    0.0    0.0    1.0    1.0
   4.0    4.0    3.0    3.0    3.0   19.0   22.0   19.0    9.0    9.0
   4.0    2.0   13.0
```

```
2 :   2  2  2  2  2  2  1  1  1  1  1  1  1  1  1
      2  2  2  2  2  2  2  2

4 :   6  6  6  6  6  6  3  3  3  3  4  4  4  4  4
      5  5  5  6  6  6  6  6

8 :  13 13 13 13 13 13  7  7  8  7  9  9 10 10  9
     11 12 11 14 14 13 13 14


      7  7  7  7  7  7  1  1  2  1  3  3  4  4  3
      5  6  5  8  8  7  7  8


     24.16        26.85        27.53        3.86        4.92
      7.45        18.90
```

Fig. E 11.1

```
M= 59   L= 2   NMAX=33   IDR=1


  54.0     0.0  -31.0    8.0    1.0  -36.0  -22.0    0.0   34.0   28.0
  12.0   -21.0   -6.0   21.0   38.0  -24.0  -38.0   86.0   58.0   -9.0
  70.0   -20.0  -43.0   59.0   -5.0   83.0   27.0   12.0   30.0   31.0
 -57.0    44.0    7.0   54.0   65.0  -35.0   46.0    5.0   56.0  -21.0
 -40.0   -43.0   57.0    0.0   25.0   56.0  -34.0  -24.0  -25.0   64.0
  63.0    37.0   -5.0    2.0  -18.0  -10.0   12.0  -40.0  -16.0
 -65.0    71.0   53.0  111.0   -9.0   52.0  -76.0   20.0  129.0   84.0
 -38.0   -26.0  -41.0   45.0  -90.0   10.0   35.0  -57.0   -1.0   -3.0
 -74.0    70.0   44.0  -26.0  114.0  -41.0  153.0  -49.0  -65.0  -12.0
  28.0   -28.0   -7.0   -8.0   -8.0   25.0   79.0  118.0    4.0   54.0
  45.0    51.0  -21.0    0.0   15.0  -25.0   56.0   36.0   49.0  -26.0
 -48.0   155.0  -24.0   28.0  -58.0   82.0  -58.0  -28.0   28.0


2 :   1  2  2  2  1  2  1  1  2  2  1  1  1  2  1
      2  2  1  1  1  1  2  2  1  2  1  2  1  1  1
      2  1  1  1  1  2  2  2  1  2  2  2  1  1  1
      1  2  2  2  1  1  2  1  2  1  2  1  1  2

4 :   4  5  5  6  3  5  3  3  6  6  3  3  3  5  4
      5  5  4  4  3  4  5  5  4  6  4  6  3  4  3
      5  4  3  4  4  5  6  6  4  5  5  5  4  3  3
      4  5  5  5  4  4  6  3  5  3  5  3  3  5

8 :   9 12 11 14  8 11  7  8 14 13  7  8  7 12  9
     11 11  9 10  8  9 12 11 10 14 10 14  7  9  8
     11 10  8 10 10 11 13 14 10 11 11 11 10  8  8
     10 11 11 11 10  9 14  8 11  7 12  7  7 11


      8 14 12  2  5 12  3  5  2  1  3  6  3 13  7
     11 12  8 10  5  8 14 12 10  2  9  2  3  7  5
     12 10  5 10 10 11  1  2 10 12 12 12 10  5  5
     10 12 12 12 10  8  2  6 11  3 14  3  4 11

     104.98       60.98        86.65        45.79        48.58
      39.72       52.21        45.95        33.00        38.02
      36.46       28.81        42.68
```

Fig. E 11.2

The 59 towns are considered again in the second example, Fig. E11.2. Fig. B31 shows the corresponding binary tree, the numbering of each level, the relevant centroid distances, and the changed numbering of the final division obtained. Again the corresponding division of the towns is displayed, in Fig. B32, from which the formation of the binary tree is clearly recognisable.

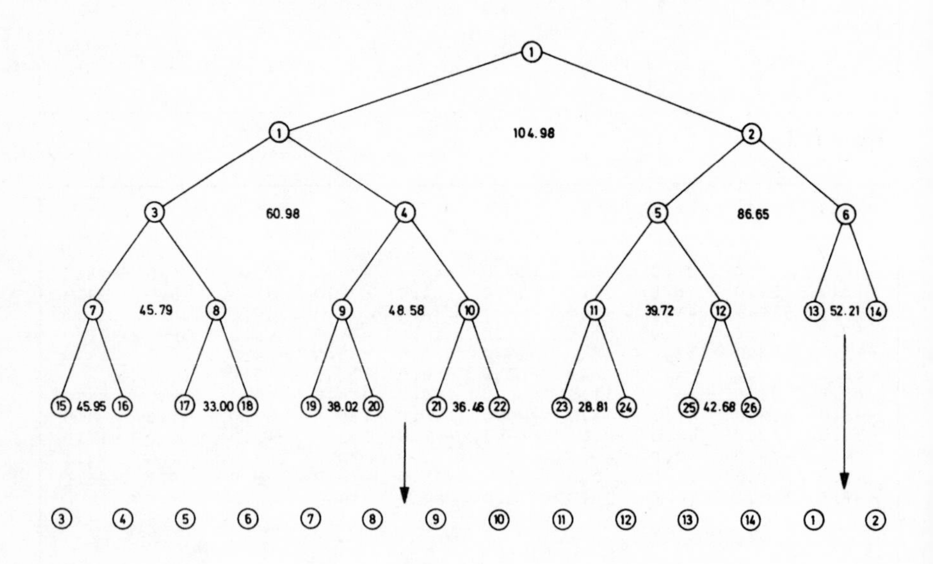

Fig. B 31

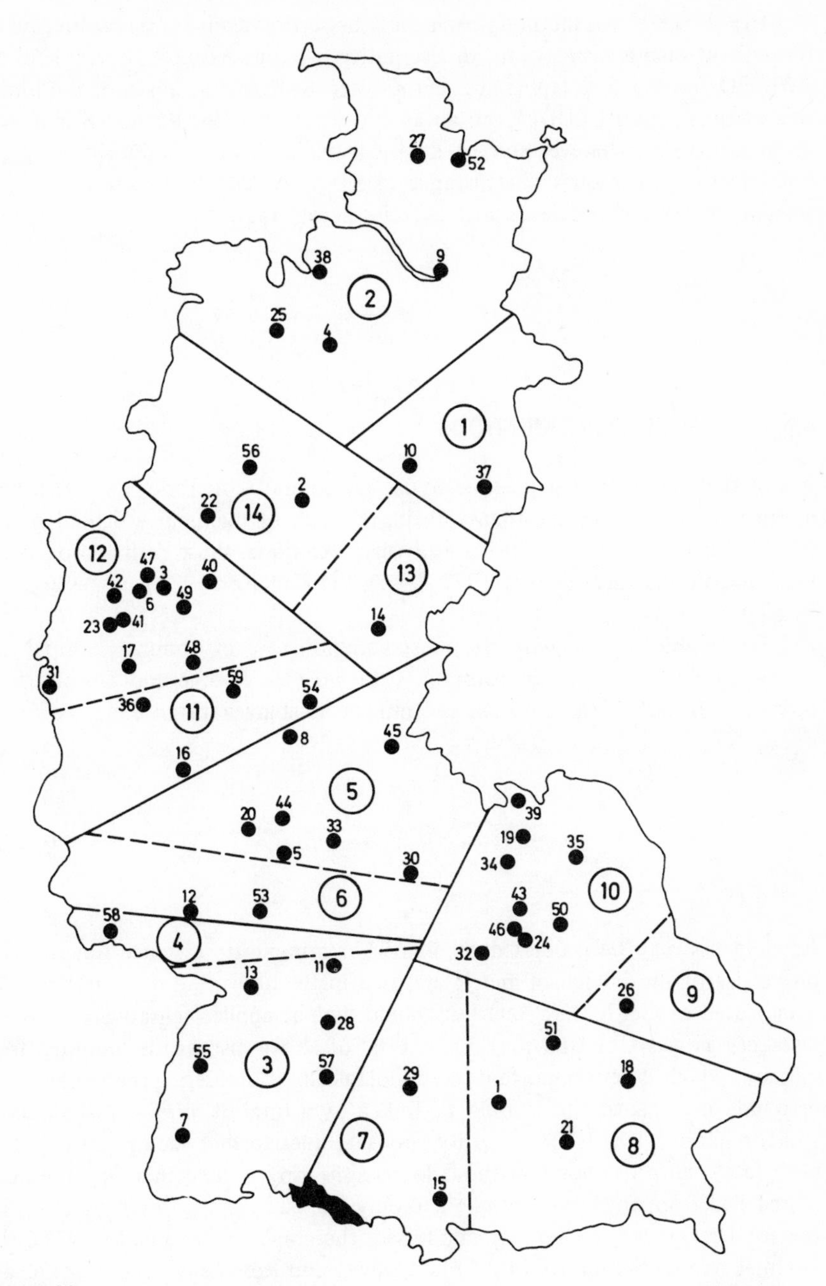

Fig. B 32

Use of these two methods could only be recommended if one really wished to perform such a strategy. As an alternative it would have been possible to use ZWEIGO for the first (splitting) method and KMEANS in place of the binary tree method. Since CLUDIA can always be substituted for KMEANS in a given distance matrix, whatever sort of data it is made up of, it is possible to apply both division techniques if a distance matrix provided for the objects, so that the limitation to metric data matrices is not necessary.

4.2 AGGLOMERATIVE METHODS

In just the same way it is possible to use agglomerative methods for metric data matrices. However, rather than splitting clusters of maximal weighted or unweighted centroid distances using a divisive technique, those with appropriately low centroid distances (Lessig 1972, Lessig and Tollefson 1971) are merged step by step.

To be able to deal with the most common case of nominal, ordinal and mixed data where one can form no centroid (see Jardine and Sibson 1971, Sokal and Sneath 1963) we shall continue to use a given distance matrix for the m objects

$$d_{ij} \geqq 0 \quad (i, j = 1, \ldots, m) \tag{4.2.1}$$

Its elements may be generated by metric or non-metric distance functions. If properties of the Euclidean metric are used in the derivation of a method, this is indicated explicitly. Such methods could then be applied tentatively for other distances as was CLUDIA previously. First of all we discuss the 'minimal tree' method, which is very important methodologically. Considering the m objects as m points in a plane, these could be linked by a total of $m(m-1)/2$ directed straight lines of length $d_{ij} = d_{ji}$. By choosing lines so that each point is linked to at least one other point, without loops appearing, and so that any point can be reached from any other by one and only one path, and if this is done so that the total length of the lines is minimised, the graph thus defined is called the minimal tree of the matrix (4.2.1) (see Gower and Ross 1969); the m points are linked by $m-1$ lines with minimum total length. Fig. B33 shows the minimal tree for the 22 towns, based on the Euclidean distances between the pairs of coordinates.

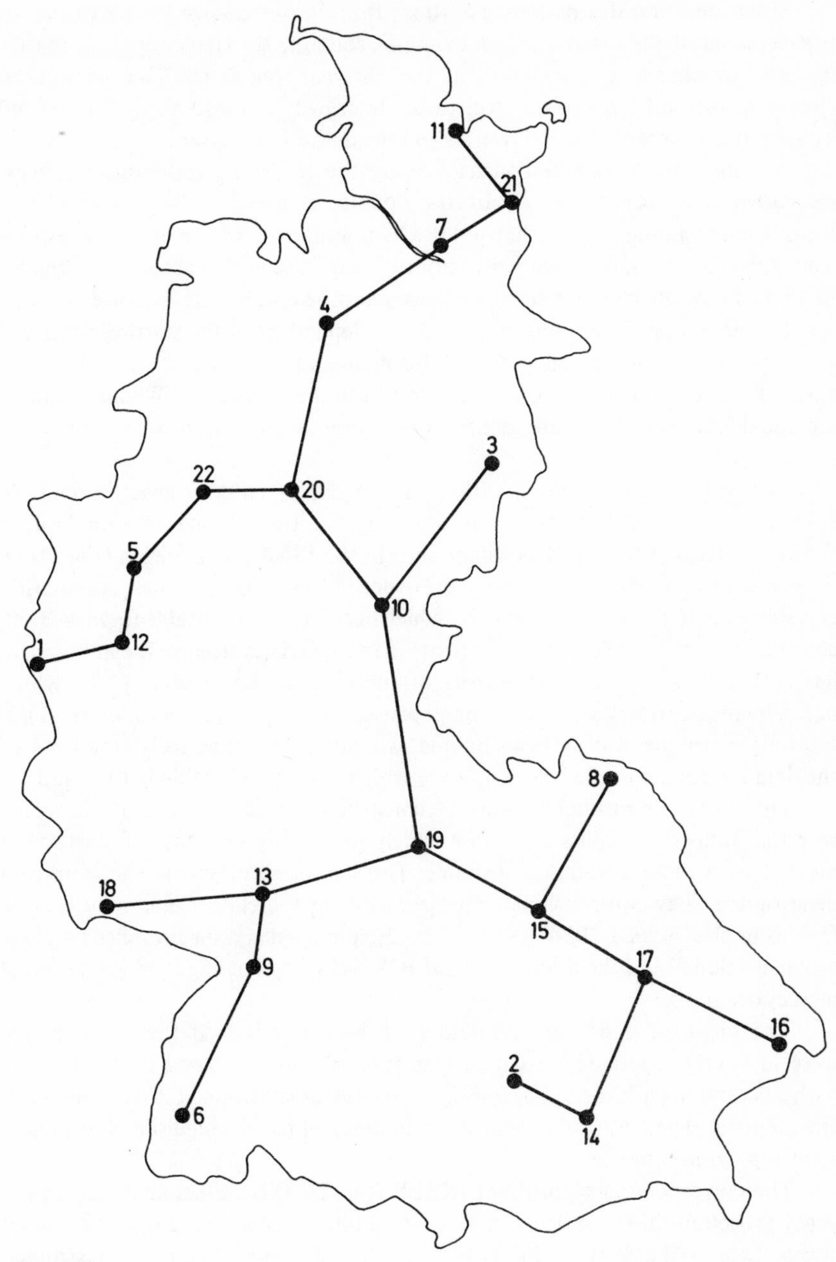

Fig. B 33

Monotonic transformations, i.e. those that do not change the rank order of the elements of the matrix, as, for example, squaring the elements of the matrix $d_{ij.}$ do not change the structure of the minimal tree at all. The hierarchical clustering method based on it, still to be described, will also not be influenced by such transformations which were also referred to in Chapter 2.

A minimal tree can be constructed by the following very simple process (see Gower and Ross 1969). An arbitrary point i is linked to the nearest point j. From the remaining points, that point k is then linked with i or with j respectively (and, in at a later stage, with any of the points which are already linked) so that firstly no closed subgraph appears and, secondly, the distance is mimimised. The minimal tree thus produced is independent of the starting point and is unique if the d_{ij} are all different from one another, which can always be ensured by introducing insignificant modifications to the d_{ij}. Where distances are equal, one or another linkage has to be chosen, which means that the uniqueness is lost.

The way in which the algorithm can be implemented is given in detail in (Gower and Ross 1969). The main advantage of the minimal tree method, as shown in (Ross 1969a), and as will be seen in the LINKER algorithm (Fig. B17), is that each element of the matrix d_{ij} is needed just once. Initial computation and subsequent storage of the matrix, which becomes impracticable for $m \approx 1000$, can therefore be avoided; if an element of the matrix is needed it can be calculated directly from the row vectors of the original data matrix (1.1), which normally requires markedly fewer storage places – namely $m.l$ instead of $m(m-1)/2$. In contrast to the KMEANS technique, which works sequentially, the rows of the data matrix are needed in a sequence which is not predictable beforehand.

The cluster algorithm based on the constructed minimal tree works in such a way that individual points and those which are already elements of clusters are merged with those at minimal distance. This can be displayed in the form of a dendrogram, a two-dimensional representation of the cluster hierarchy created (see examples in Figs. B34 to B40). In these diagrams a measurement scale is given additionally to the branching trees B29 and B31, showing at which distances merges occur.

The identical dendrogram would have been obtained if the minimal tree were divisively separated, in which case the divisions are based on the clusters with maximum separation. Accordingly, the agglomerative and divisive methods are identical when the minimal tree is unique, which distinguishes this cluster algorithm from others.

The corresponding algorithm LINKER (Fig. U17) is a Fortran version of the Algol procedure in (Ross 1969a), with the addition of a sort. From the output arrays Q(M – 1), P(M – 1), T(M – 1) the dendrogram can be constructed unambiguously in the following way. At step i ($i = 1, \ldots, m-1$) the clusters with the identifying numbers Q(I) and P(I) with minimal distance T(I) are merged. In this, Q(I) and P(I) respectively are to be interpreted so that this cluster number is some object number of those objects which have already been merged. The monotonicity of T(I) can then be used to reconstruct accurately which object numbers belong together at which stage. (Anderberg 1973, Gower

Fig. U 17

```
      SUBROUTINE LINKER (M,D,Q,P,T)
C
C     CONSTRUCTION OF THE MINIMAL TREE OF A SYMMETRICAL MATRIX
C     D(M,M).
C
C     FOR EACH I=1,..,M-1 A PARTNER POINT P(I) NOT EQUAL TO I
C     (1 <= P(I) <= M) IS SOUGHT, AT MINIMUM DISTANCE T(I), SUCH
C     THAT ALL POINTS ARE SIMPLY CONNECTED.  THE TREE HAS
C     MINIMUM TOTAL LENGTH (I.E. IS THE MINIMAL SPANNING TREE).
C
C     THE TRIPLES (I,P(I),T(I)) ARE ORDERED ACCORDING TO THE
C     MAGNITUDE OF T(I) AND THE ORDERED TRIPLES ARE STORED
C     IN (Q(I),P(I),T(I)).
C
C     THE TWO POSSIBLE HIERARCHICAL CLUSTER PROCESSES -
C     DIVISIVE (SPLITTING WHERE T(I) IS LARGEST) - AND
C     AGGLOMERATIVE (MERGING WHERE T(I) IS SMALLEST) - ARE
C     IDENTICAL.  THE DENDROGRAM MAY BE CONSTRUCTED WITH THE
C     AID OF Q, P, AND T.
C
C     IN THE ALGORITHM EACH D(I,J) IS USED ONLY ONCE AND COULD
C     THEREFORE BE CALCULATED AT THIS POINT ONLY.
C
C     ONLY THE UPPER TRIANGULAR PART OF THE MATRIX D IS USED.
C
C     DIMENSION D(M,M),  P(M-1),T(M-1),Q(M-1)
      DIMENSION D(60,60),P(60), T(60), Q(60)
      INTEGER P,Q,R,S
      LOGICAL ENDE
      IF(M.LE.1) RETURN
      M1=M-1
      V=1.E30
      DO 1 K=1,M1
           Q(K)=0
           P(K)=0
           T(K)=V
    1 CONTINUE
      J=M
      DO 4 I=1,M1
           U=V
           DO 3 K=1,M1
                IF(Q(K).NE.0) GOTO 3
                R=MIN0(J,K)
                S=MAX0(J,K)
                H=D(R,S)
                IF(H.GE.T(K)) GOTO 2
                T(K)=H
                P(K)=J
    2           IF(T(K).GE.U) GOTO 3
                U=T(K)
                N=K
```

```
3       CONTINUE
        J=N
        Q(J)=1
4  CONTINUE
   DO 5 I=1,M1
        Q(I)=I
5  CCNTINUE
   ENDE=.FALSE.
   DO 7 I=2,M1
        IF(ENDE) RETURN
        ENDE=.TRUE.
        LL=M1-I+1
        DO 6 L=1,LL
             N=L+1
             IF(T(L).LE.T(N)) GOTO 6
             V=T(N)
             T(N)=T(L)
             T(L)=V
             J=Q(N)
             Q(N)=Q(L)
             Q(L)=J
             J=P(N)
             P(N)=P(L)
             P(L)=J
             ENDE=.FALSE.
6       CONTINUE
7  CONTINUE
   END
```

Fig. U 17

and Ross 1969, Ross 1969c, Rohlf 1974) give Algol and Fortran algorithms to produce representations of the dendrograms automatically, which we have not undertaken here. Other Fortran programs for the minimal tree method, which we do not consider to be any better, are put forward in (Jardine and Sibson 1971, Rijsbergen 1970, Sale 1971, Seppanen and Sibson 1973). The subroutine HIERCL, which will be described in the following examples, uses the minimal tree method in yet another form (see Johnson 1963).

In main program H12, if IT$\neq$0 the elements of the upper triangular matrix d_{ij} that are read in are squared before being used.

Fig. E12.1 makes use of a matrix of rank orders for pairs of cars (Green and Tull 1970), the upper half of which is printed in rows. For example $d_{36} = 1$ means that the cars numbered 3 and 6 had the least separation and $d_{35} = 55 = 11(11-1)/2$ means that those with the numbers 3 and 5 had the greatest separation. Fig. B34 shows the dendrogram corresponding to the values Q(I), P(I) and T(I).

```
C
C      SINGLE LINKAGE CLUSTER ANALYSIS
C
       DIMENSION D(60,60),P(60),Q(60),T(60)
       INTEGER P,Q
       KI=5
       KO=6
     1 READ(KI,2) M,K,K,K,K,IT
     2 FORMAT(16I5)
       IF(M.LE.2.OR.M.GT.60) STOP
       WRITE(KO,3) M,IT
     3 FORMAT('1',' M=',I2,' IT=',I1)
       WRITE(KO,4)
     4 FORMAT('0')
       M1=M-1
       DO 6 I=1,M1
            I1=I+1
            READ(KI,7)  (D(I,J),J=I1,M)
            WRITE(KO,8) (D(I,J),J=I1,M)
            IF(IT.EQ.0) GOTO 6
            DO 5 J=I1,M
                 H=D(I,J)
                 D(I,J)=H*H
     5      CONTINUE
     6 CONTINUE
     7 FORMAT(16F5.0)
     8 FORMAT(1X,7F9.2)
       CALL LINKER(M,D,Q,P,T)
       WRITE(KO,4)
       DO 9 I=1,M1
            WRITE(KO,10) Q(I),P(I),T(I)
     9 CONTINUE
    10 FORMAT(1X,2I4,F10.2)
       GOTO 1
       END
```

Fig. H 12

```
M=11   IT=0

    8.00     50.00     31.00     12.00     48.00     36.00      2.00
    5.00     39.00     10.00
   38.00      9.00     33.00     37.00     22.00      6.00      4.00
   14.00     32.00
   11.00     55.00      1.00     23.00     46.00     41.00     17.00
   52.00
   44.00     13.00     16.00     19.00     25.00     18.00     42.00
   54.00     53.00     30.00     28.00     45.00      7.00
   26.00     47.00     40.00     24.00     51.00
   29.00     35.00     34.00     49.00
    3.00     27.00     15.00
   20.00     21.00
   43.00

  6   3      1.00
  8   1      2.00
  9   8      3.00
  2   9      4.00
  5  11      7.00
  4   2      9.00
  1  11     10.00
  3   4     11.00
 10   2     14.00
  7   4     16.00
```

Fig. E 12.1

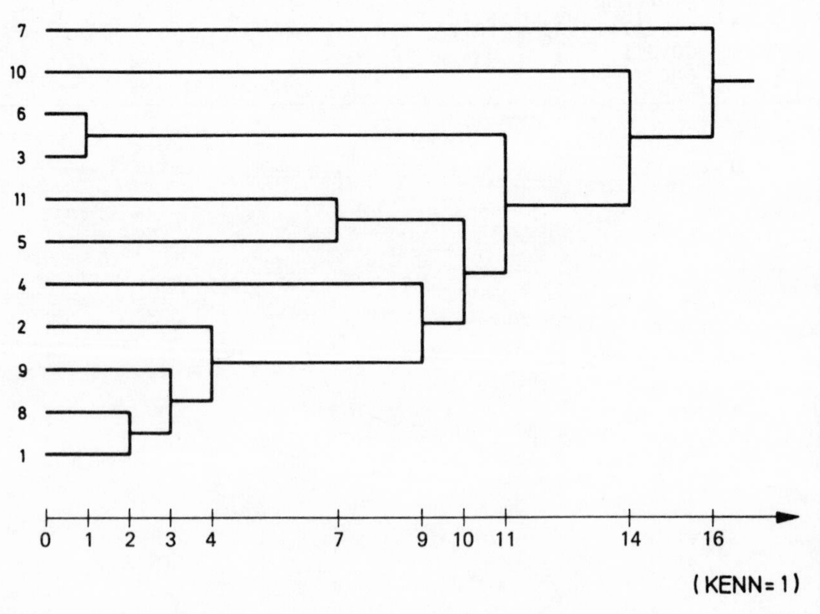

Fig. B 34

Fig. E12.2 uses a matrix of rank orders for 15 different types of breakfast, derived in a suitable way from matrices of rank orders of a number of people (see Green and Rao 1969). The dendrogram corresponding to the output will be found later, in Fig. B37.

```
M=15   IT=0


  59.13     62.42     43.64     36.01     60.21     78.01     34.57
  32.10     61.52     51.26     46.65     55.17     53.67     72.17
  30.79     83.89     44.15     60.94     27.02     36.36     32.29
   8.49     78.35     84.55     83.82     83.01     48.13
  82.37     57.76     23.80     26.85     52.44     50.18     25.29
  72.21     76.14     78.07     70.68     21.87
  65.46     53.33     93.71     55.35     49.11     85.64     32.68
  21.40      7.83     30.56     73.62
  64.11     72.92     32.87     36.29     48.96     21.65     49.30
  55.70     49.87     62.49
  50.49     59.69     55.83     58.73     47.42     46.98     55.00
  47.08     15.25
  64.23     61.55     31.52     78.21     83.70     88.85     78.13
  33.37
   5.52     37.94     64.23     59.80     60.38     63.77     63.52
  38.32     61.30     57.48     65.85     65.56     64.49
  80.14     86.42     83.86     84.24     52.93
  20.31     23.33     23.10     58.26
  20.54     10.70     62.46
  24.76     66.77
  54.17


  8    9      5.52
  4   13      7.83
  2   10      8.49
 14   12     10.70
  6   15     15.25
 12   11     20.31
 13   12     20.54
 11    5     21.65
  3   15     21.87
 10    3     25.29
  7    3     26.85
  1    9     32.10
  9    2     32.29
  5    8     32.87
```

Fig. E 12.2

There is an entire class of agglomerative hierarchical clustering methods (Cunningham and Ogilvie 1972, Duda and Hart 1973), in which the two objects or two clusters C_p and C_q are merged whose separation is minimal. Thus with

$$d_{pq} = \min_{i>j} d_{ij}, \qquad (4.2.2)$$

the cluster produced $C_k = C_p \cup C_q$, and the distances to all other clusters C_i are redefined by means of

$$d_{ki} = \alpha_p d_{pi} + \alpha_q d_{qi} + \beta d_{pq} - \gamma |d_{pi} - d_{qi}| \quad (i \neq p, q) \qquad (4.2.3)$$

α_p, α_q, β and γ are constants which ultimately depend on the size of the cluster. (Lance and Williams 1967a) suggest choosing $\gamma = 0$, $\beta < 0$ and $\alpha_p + \alpha_q + \beta \geqslant 1$; (Rohlf 1970) suggests continually adapting the coefficients to the degree of merging.

We shall now give seven particular methods of type (4.2.3) (see Cunningham and Ogilvie 1972); describe a unified program which uses them; discuss the results of various examples; and make recommendations based on these results which agree with those quoted in the literature. The first two methods are

$$d_{ki} = \frac{1}{2}(d_{pi} + d_{qi}) - \frac{1}{2}|d_{pi} - d_{qi}| \qquad (4.2.4)$$

$$= \min(d_{pi}, d_{qi})$$

and

$$d_{ki} = \frac{1}{2}(d_{pi} + d_{qi}) + \frac{1}{2}|d_{pi} - d_{qi}| \qquad (4.2.5)$$

$$= \max(d_{pi}, d_{qi}).$$

(Johnson 1967). In (4.2.4) the distances from $C_k = C_p \cup C_q$ to C_i were replaced by the minimum of the distances from C_p to C_i and C_q to C_i respectively and in (4.2.5) by the maximum. It can be shown that method (4.2.5) is identical to the minimal tree method discussed earlier. Admittedly it looks as if it were quite different, but put in graph theoretic terms (4.2.5) corresponds to dividing a tree formed by linking firstly the two points with the least separation, then those with the next-to-least separation, etc., and finally those with the greatest separation (see Anderberg 1973).

(Sokal and Sneath 1963) recommend taking the average

$$d_{ki} = \frac{1}{2}(d_{pi} + d_{qi}) \qquad (4.2.6)$$

as being the method which is at once the simplest and the most successful.

The method

$$d_{ki} = \frac{1}{2}(d_{pi} + d_{qi}) - \frac{1}{4}d_{pq} \qquad (4.2.7)$$

(see Gower 1967) is a special case of (4.2.9) below and

$$d_{ki} = \frac{m_p d_{pi} + m_q d_{qi}}{m_p + m_q} \qquad (4.2.8)$$

is a generalisation of (4.2.6) using the number of objects as weights.

The sixth method (Gower 1967, Bock 1970, Bock 1974) is

$$d_{ki} = \frac{m_p}{m} d_{pi} + \frac{m_q}{m} d_{qi} - \frac{m_p m_q}{m^2} d_{pq} \quad (m = m_p + m_q) \tag{4.2.9}$$

which is derived from

$$d_{pq} = d_2^2(\bar{x}_p, \bar{x}_q) = \|\bar{x}_p - \bar{x}_q\|^2 \tag{4.2.10}$$

by the use of

$$m_k \bar{x}_k = m_p \bar{x}_p + m_q \bar{x}_q \quad (m_k = m)$$

and the conversion of (4.1.4) into (4.1.5), as follows:

$$\begin{aligned}
&\frac{m_p}{m} \|\bar{x}_p - \bar{x}_i\|^2 + \frac{m_q}{m} \|\bar{x}_q - \bar{x}_i\|^2 - \frac{m_p m_q}{m^2} \|\bar{x}_p - \bar{x}_q\|^2 \\
&= \frac{m_p}{m} \|(\bar{x}_p - \bar{x}_k) + (\bar{x}_k - x_i)\|^2 + \frac{m_q}{m} \|(\bar{x}_q - \bar{x}_k) + (\bar{x}_k - \bar{x}_i)\|^2 \\
&\quad - \left(\frac{m_p}{m} \|\bar{x}_p - \bar{x}_k\|^2 + \frac{m_q}{m} \|\bar{x}_q - \bar{x}_k\|^2\right) \\
&= \|x_k - x_i\|^2 .
\end{aligned}$$

We must stress that (4.2.9) only applies if the squared Euclidean metric is used, and accordingly presupposes a metric data matrix. Despite this, (4.2.9) can be used for any kind of distance matrix. Method (4.2.7) results from (4.2.9) when the weights are equal.

By defining the distance as

$$d_{pq} = \frac{m_p m_q}{m_p + m_q} \|\bar{x}_p - \bar{x}_q\|^2 , \tag{4.2.11}$$

instead of (4.2.10), and multiplying both sides of (4.2.9) by

$$\frac{(m_p + m_q)\, m_i}{m_p + m_q + m_i}$$

and making several conversions on the right-hand side, the method

$$d_{ki} = \frac{1}{m_p + m_q + m_i} \left[(m_p + m_i)\, d_{pi} + (m_q + m_i)\, d_{qi} - m_i\, d_{pq}\right]. \tag{4.2.12}$$

is obtained (see Ward 1963). Since, because of (3.2.4),

$$d_{pq} = \sum_{j \in C_k} \|x_j - \bar{x}_k\|^2 - \sum_{j \in C_p} \|x_j - \bar{x}_p\|^2 - \sum_{j \in C_q} \|x_j - \bar{x}_q\|^2, \quad (4.2.13)$$

holds for (4.2.11), at each merge the loss of information is minimised, as long as minimal (4.2.13) is selected, which is characteristic of this method. A further Fortran algorithm for it is given in (Veldman 1967). The same remarks hold for (4.2.12), that is, that metric data with Euclidean distances is required, and the range of possible applications, as were made for (4.2.9). For example, (Kernan and Bruce 1972) use Mahalanobis distances in (4.2.12).

All seven methods (4.2.4), (4.2.5), (4.2.6), (4.2.7), (4.2.8), (4.2.9) and (4.2.12) are implemented in parallel in the subroutine HIERCL (Fig. U18), for KENN = 1, . . . ,7. In contrast with LINKER, when HIERCL is called the upper triangular matrix d_{ij} ($j > i \geqslant 1$) is overwritten. The developing dendrogram is again characterised by three arrays A(M – 1), B(M – 1) and H(M – 1), which roughly correspond to Q, P, and T in LINKER. Reconstructing the dendrogram is somewhat simpler here, because B(I) contains only cluster numbers that represent single objects or values A(J) for J < I.

Fig. U 18

```
      SUBROUTINE HIERCL (M,D,KENN,A,B,H)
C
C     SEVEN AGGLOMERATIVE HIERARCHICAL METHODS.
C
C     ONE OF THE SEVEN ALTERNATIVE METHODS, SPECIFIED BY THE
C     VALUE OF KENN = 1,...,7, IS APPLIED TO THE SYMMETRIC
C     MATRIX D(M,M).
C
C     THE NUMBERS OF THE OBJECTS OR CLUSTERS MERGED AT LEVEL K
C     ARE PLACED IN THE ARRAYS A AND B, B(K) BEING MERGED WITH
C     A(K), THE SEPARATION BEING H(K).    IT IS POSSIBLE FROM A,
C     B, AND H TO CONSTRUCT A DENDROGRAM.
C
C     ONLY THE UPPER TRIANGULAR PART OF D IS USED, BUT IT IS
C     DESTROYED.
C
C     DIMENSION D(M,M)   ,A(M-1),B(M-1),H(M-1)
      DIMENSION D(60,60),A(60), B(60), H(60)
C     DIMENSION P(M),Q(M)
      DIMENSION P(60),Q(60)
      INTEGER   A,B,P,Q,QIC,QJC,QI
      IF(KENN.LT.1.OR.KENN.GT.7) RETURN
      M1=M-1
      K=0
      DO 1 I=1,M
           P(I)=0
           Q(I)=1
```

```
      1 CONTINUE
      2 K=K+1
        IF(K.EQ.M) RETURN
        DMAX=1.E30
        DO 4 I=1,M1
              IF(P(I).NE.0) GOTO 4
              I1=I+1
              DO 3 J=I1,M
                    IF(P(J).NE.0) GOTO 3
                    T=D(I,J)
                    IF(T.GT.DMAX) GOTO 3
                    IC=I
                    JC=J
                    DMAX=T
      3       CONTINUE
      4 CONTINUE
        P(JC)=1
        A(K)=IC
        B(K)=JC
        H(K)=DMAX
        IF(KENN.LT.5) GOTO 5
        QIC=Q(IC)
        QJC=Q(JC)
        IF(KENN.NE.7) F=1./FLOAT(QIC+QJC)
      5 DO 20 I=1,M
              IF(I.EQ.IC.OR.P(I).NE.0) GOTO 20
              IF(KENN.NE.7) GOTO 6
              QI=Q(I)
              F=1./FLOAT(QI+QIC+QJC)
      6       J=MIN0(IC,I)
              L=MAX0(IC,I)
              K1=MIN0(JC,I)
              K2=MAX0(JC,I)
              DJ=D(J,L)
              DK=D(K1,K2)
              GOTO (11,12,13,14,15,16,17),KENN
C
     11       D(J,L)=AMIN1(DJ,DK)
                                                                     GOTO 20
     12       D(J,L)=AMAX1(DJ,DK)
                                                                     GOTO 20
     13       D(J,L)=.5*(DJ+DK)
                                                                     GOTO 20
     14       D(J,L)=.5*(DJ+DK)-.25*DMAX
                                                                     GOTO 20
     15       D(J,L)=F*(QIC*DJ+QJC*DK)
                                                                     GOTO 20
     16       D(J,L)=F*(QIC*DJ+QJC*DK-QIC*QJC*F*DMAX)
                                                                     GOTO 20
     17       D(J,L)=F*((QIC+QI)*DJ+(QJC+QI)*DK-QI*DMAX)
C
     20 CONTINUE
        Q(IC)=Q(IC)+Q(JC)
        GOTO 2
        END
```

Fig. U 18

In main program H13 the parameter IT has the same meaning as in H12. If a value in the range $1 \leqslant$ KENN $\leqslant 7$ is specified, only the corresponding method is used. For KENN = 0, all seven methods are applied in turn and the parameters of all seven dendrograms are printed. Since HIERCL destroys the upper triangular part of the matrix d_{ij}, it is stored before each call in the lower part, to allow it to be regenerated.

Fig. H 13

```
C
C     AGGLOMERATIVE HIERARCHICAL CLUSTERING (SEVEN METHODS).
C
C     FOR KENN = 0 ALL SEVEN METHODS ARE APPLIED.
C
C     IF 1 <= KENN <= 7, ONLY THE METHOD SO NUMBERED IS APPLIED.
C
      DIMENSION D(60,60),A(60),B(60),H(60)
      INTEGER A,B
      KI=5
      KO=6
    1 READ(KI,2) M,K,K,K,K,IT,KENN
    2 FORMAT(16I5)
      IF(M.LE.1.OR.M.GT.60.OR.KENN.LT.0.OR.KENN.GT.7) STOP
      WRITE(KO,3) M,IT,KENN
    3 FORMAT('1','  M=',I2,'   IT=',I1,'   KENN=',I1)
      WRITE(KO,4)
    4 FORMAT('0')
      M1=M-1
      IF(KENN.NE.0) GOTO 5
      K1=1
      K2=7
      GOTO 6
    5 K1=KENN
      K2=KENN
    6 DO 9 I=1,M1
         I1=I+1
         READ(KI,10)  (D(I,J),J=I1,M)
         WRITE(KO,11) (D(I,J),J=I1,M)
         DO 8 J=I1,M
            T=D(I,J)
            IF(IT.EQ.0) GOTO 7
            T=T*T
            D(I,J)=T
    7       D(J,I)=T
    8    CONTINUE
    9 CONTINUE
   10 FORMAT(16F5.0)
   11 FORMAT(1X,7F9.2)
      WRITE(KO,18)
      DO 17 KENN=K1,K2
          WRITE(KO,12) KENN
```

```
12          FORMAT('0','  KENN=',I1)
            WRITE(KO,4)
            CALL HIERCL (M,D,KENN,A,B,H)
            DO 13 I=1,M1
                  WRITE(KO,14) I,H(I),A(I),B(I)
13          CONTINUE
14          FORMAT(1X,I5,F13.2,4X,2I4)
            IF(K1.EQ.K2) GOTO 17
            DO 16 I=1,M1
                  I1=I+1
                  DO 15 J=I1,M
                        D(I,J)=D(J,I)
15                CONTINUE
16          CONTINUE
17 CONTINUE
18 FORMAT('1')
   GOTO 1
   END
```

Fig. H 13

The first example (Fig. E13.1) uses the matrix of rank orders for pairs of cars encountered earlier in Fig. E12.1, and prints the results for KENN = 1, ... ,7; the dendrograms corresponding to KENN = 1, 2, 3, 4 are shown in Figs. B34 to B37.

Fig. E 13.1

```
KENN=1

   1        1.00        3   6
   2        2.00        1   8
   3        3.00        1   9
   4        4.00        1   2
   5        7.00        5  11
   6        9.00        1   4
   7       10.00        1   5
   8       11.00        1   3
   9       14.00        1  10
  10       16.00        1   7

KENN=2

   1        1.00        3   6
   2        2.00        1   8
   3        4.00        2   9
   4        7.00        5  11
   5        8.00        1   2
   6       13.00        3   4
   7       24.00        3  10
   8       33.00        1   5
   9       34.00        3   7
  10       55.00        1   3

KENN=3

   1        1.00        3   6
   2        2.00        1   8
   3        4.00        2   9
   4        5.50        1   2
   5        7.00        5  11
   6       12.00        3   4
   7       19.25        3  10
   8       22.63        1   5
   9       27.13        3   7
  10       39.02        1   3

KENN=4

   1        1.00        3   6
   2        2.00        1   8
   3        3.50        1   9
   4        4.38        1   2
   5        7.00        5  11
   6       11.75        3   4
   7       16.19        3  10
   8       17.34        1   3
   9       19.55        1   7
  10       33.03        1   5
```

KENN=5

1	1.00	3	6
2	2.00	1	8
3	4.00	2	9
4	5.50	1	2
5	7.00	5	11
6	12.00	3	4
7	19.67	3	10
8	22.63	1	5
9	24.75	3	7
10	38.03	1	3

KENN=6

1	1.00	3	6
2	2.00	1	8
3	3.50	1	9
4	4.89	1	2
5	7.00	5	11
6	11.75	3	4
7	16.89	3	10
8	19.13	1	5
9	19.50	3	7
10	24.71	1	3

KENN=7

1	1.00	3	6
2	2.00	1	8
3	4.00	2	9
4	7.00	5	11
5	8.00	1	2
6	15.67	3	4
7	25.33	3	10
8	31.20	3	7
9	51.00	1	5
10	134.80	1	3

Fig. E 13.1

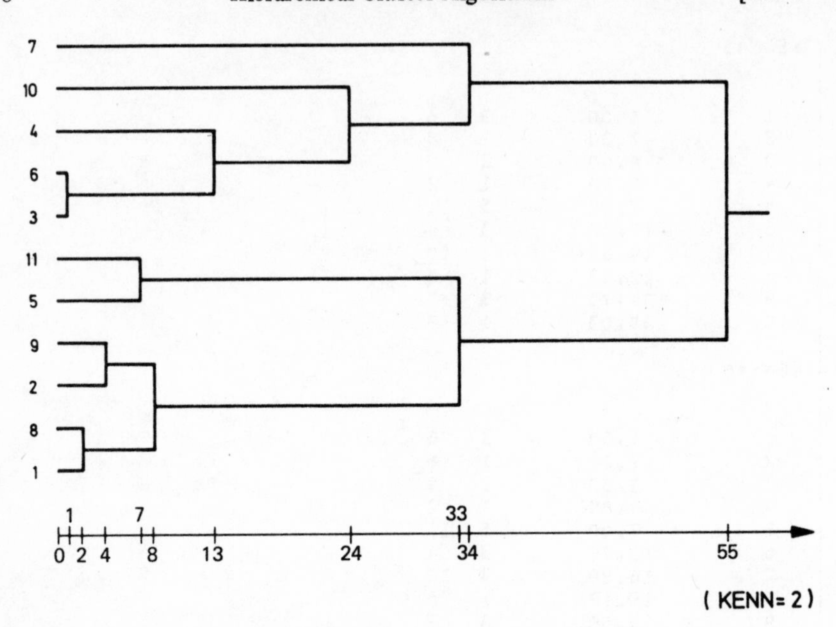

Fig. B 35

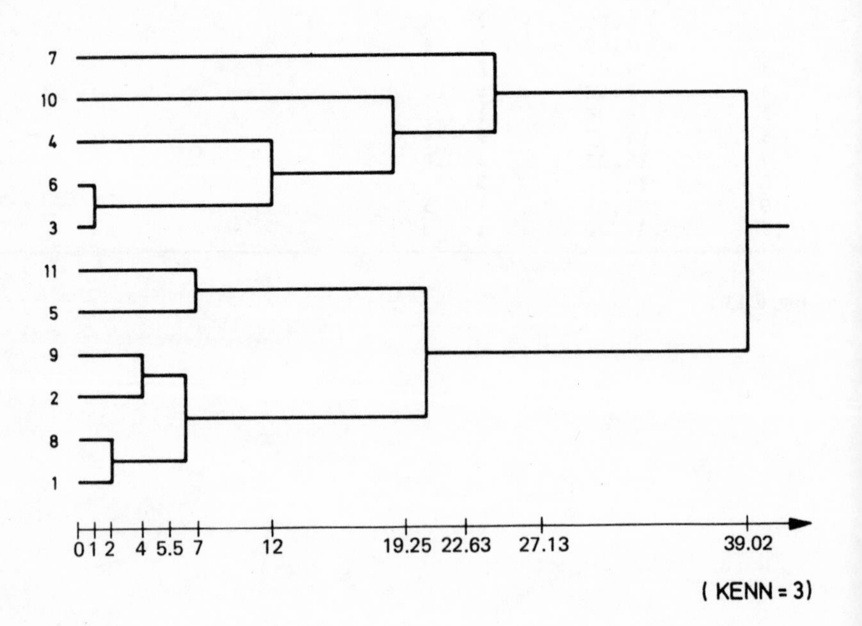

Fig. B 36

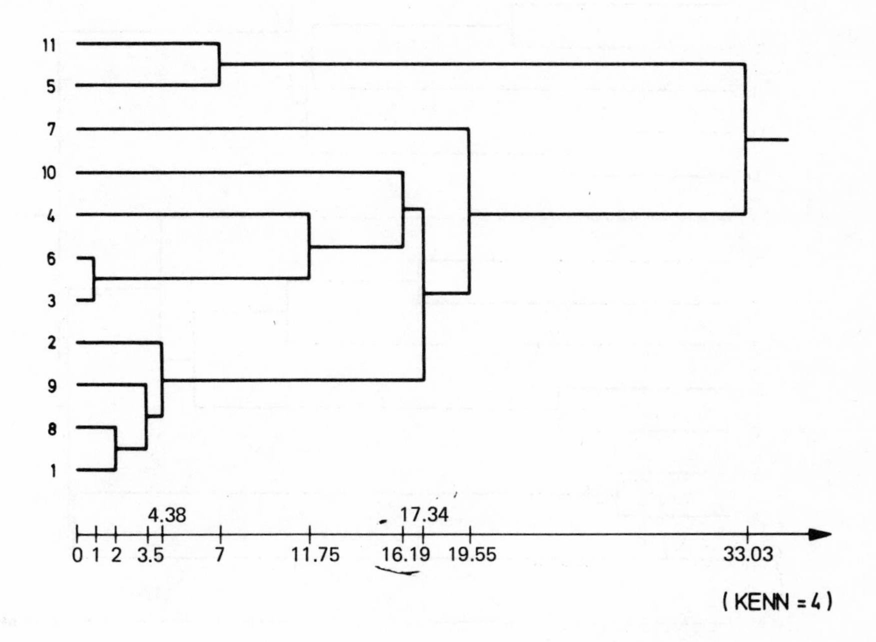

Fig. B 37

The second example uses the matrix from E12.2. Corresponding dendrograms for KENN = 1, 2, 3 are given in Figs. B38 to B40.

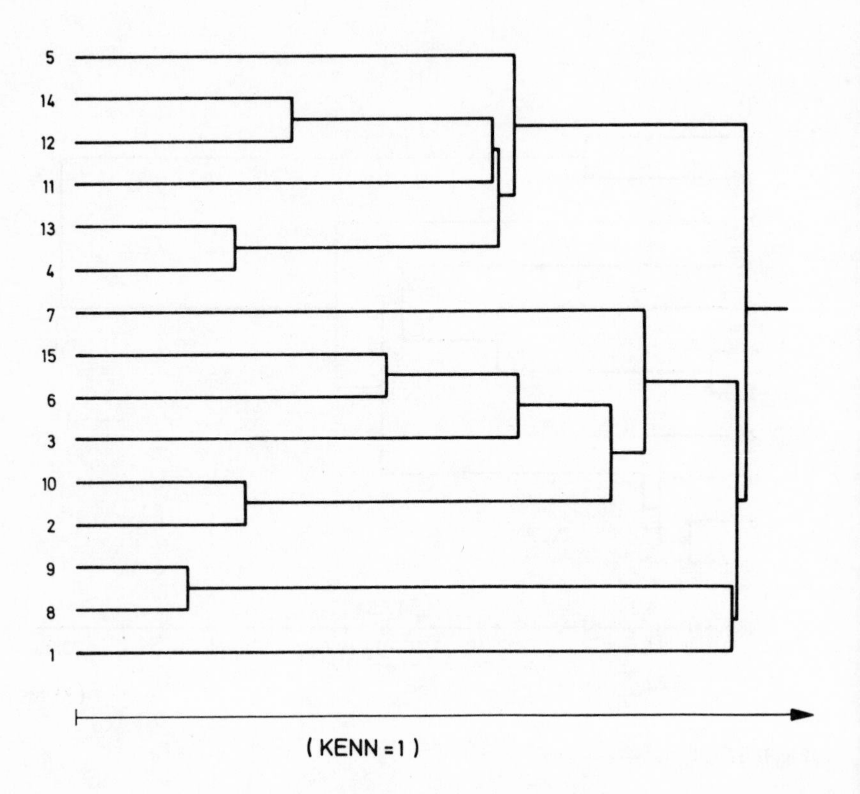

Fig. B 38

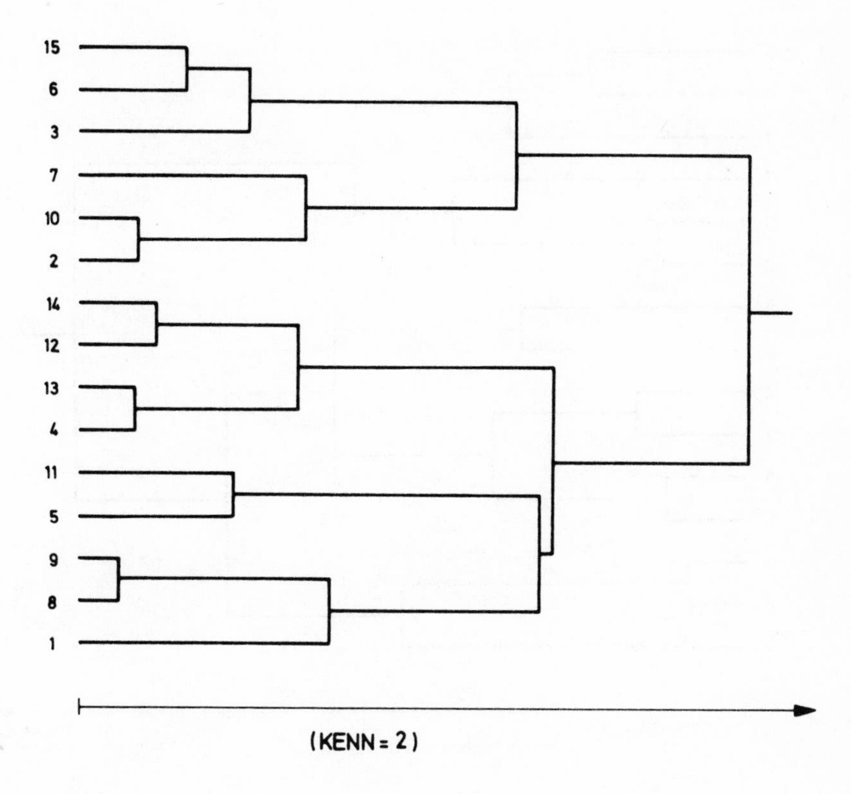

Fig. B 39

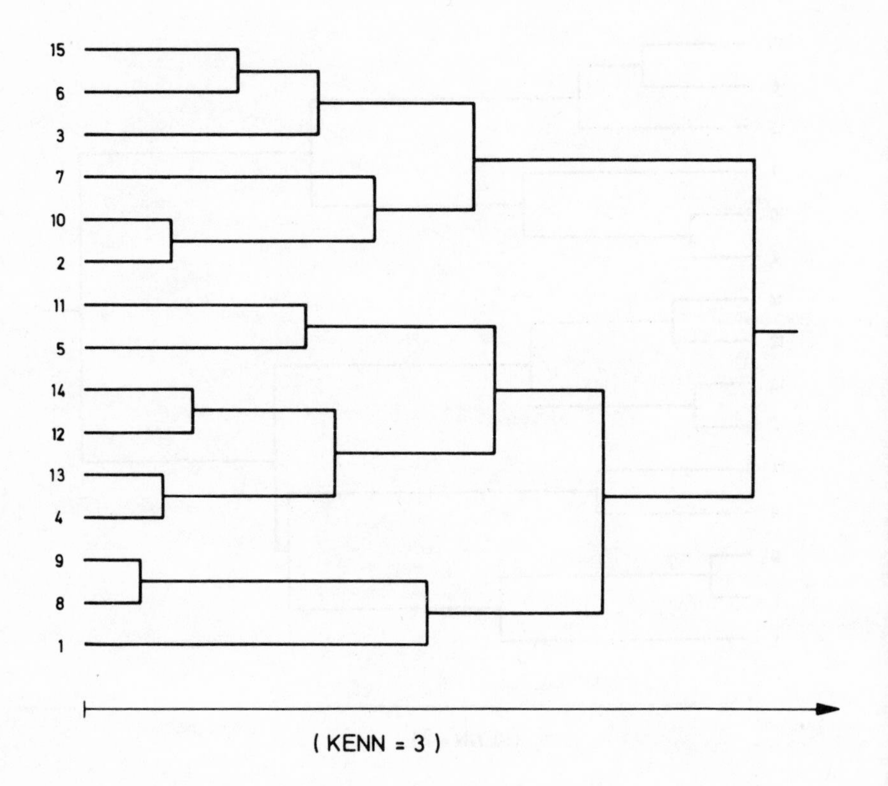

Fig. B 40

The third example uses the distance matrix for the 22 towns of earlier examples. Fig. B41 contains the resulting dendrogram for KENN = 1.

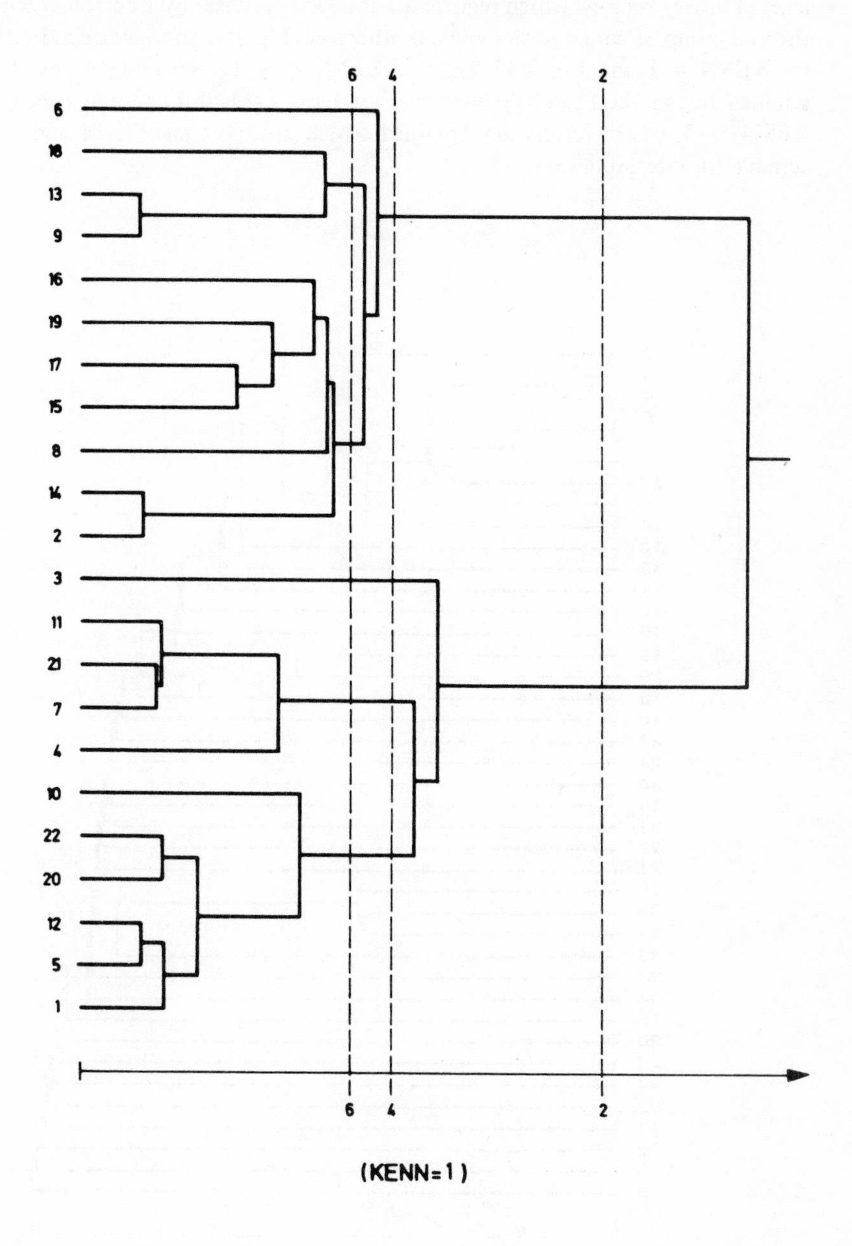

Fig. B 41

The final example uses a distance matrix for 40 kinds of commodities in a large store; this sort of matrix can be built up with the use of Tanimoto coefficients in binary vectors, which register a 1 for each purchase by a customer within the said group of commodities and a 0 otherwise. Fig. B42 shows the dendrogram for KENN = 1 and Fig. B43 that for KENN = 3. The chaining effect of the minimal tree method can be seen in the case KENN = 1; this does not appear for KENN = 3, where results are obtained which are more significant and more suitable for interpretation.

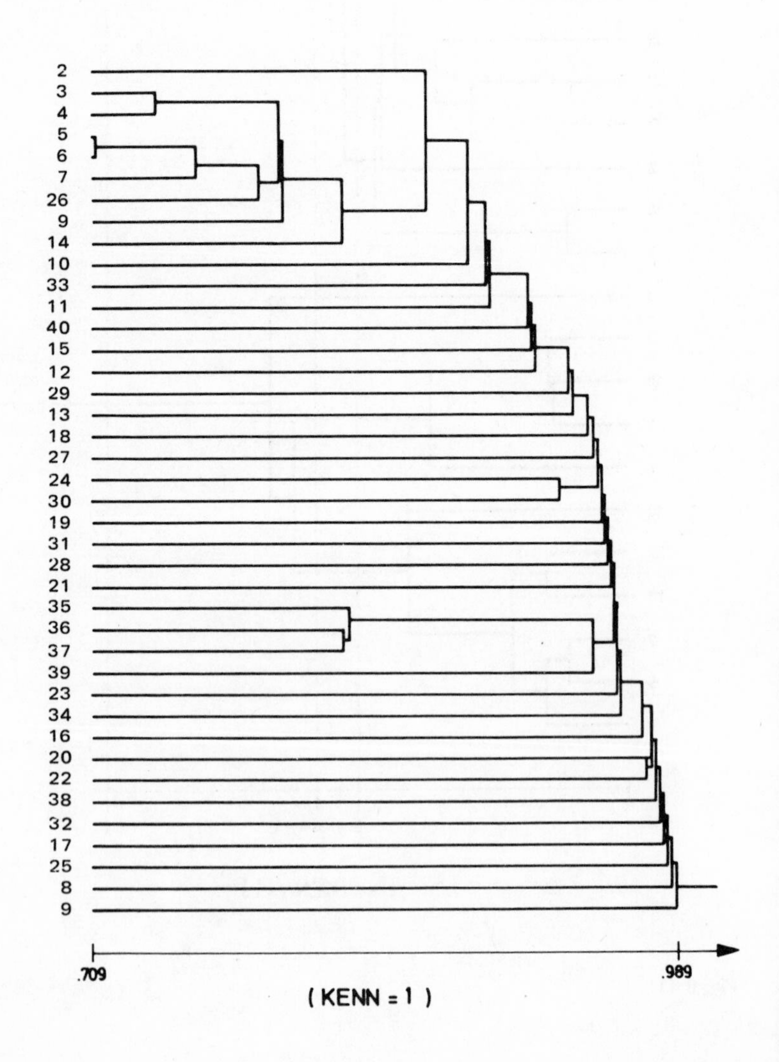

Fig. B 42

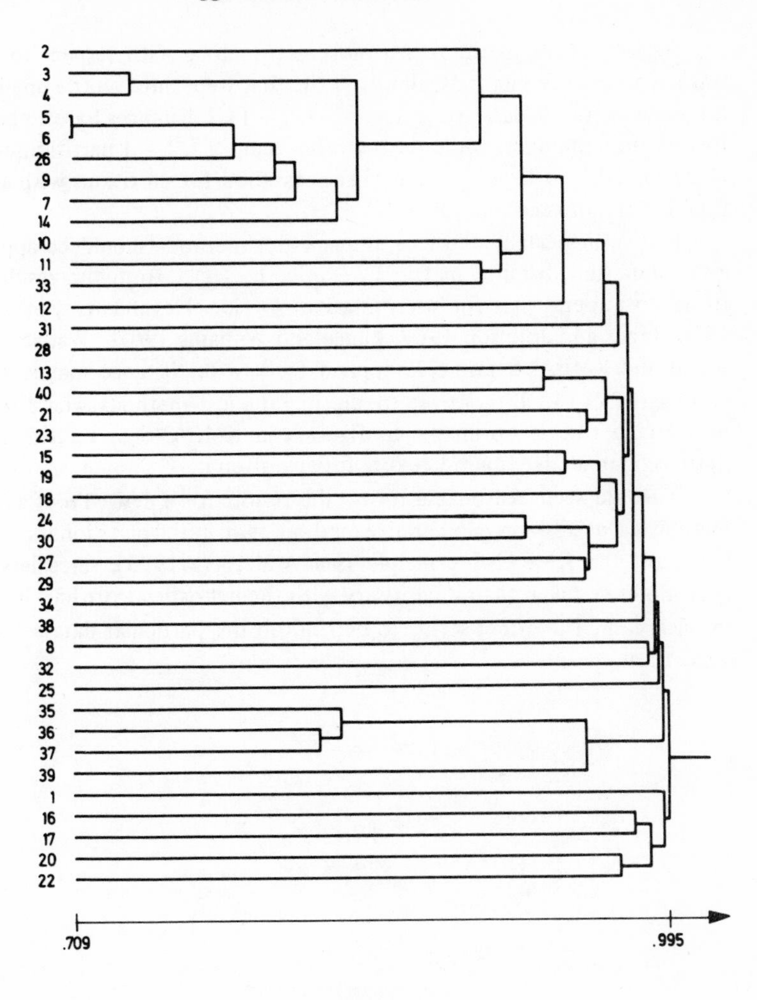

Fig. B 43

(KENN = 3)

Summarising these and the evidence from other numerical results, we recommend KENN = 7 if Euclidean distances are used and KENN = 3 otherwise. Thorough tests for the methods as a whole are included in (Cunningham and Ogilivie 1972), and evaluation criteria which give the same results in (Green and Rao 1969). A further measure to compare any sort of clustering method is presented by (Rand 1971). Agglomerative methods for correlation matrices of variables are given in (Fortier and Solomon 1966, King 1967). (Ling 1973) offers technical assistance in the interpretation of the results of hierarchical clustering methods.

Agglomerative methods are more satisfactory with respect to computing times than divisive methods, almost at the first step, since on the one hand – and the same is true of later steps too – $m(m-1)/2$ distances have to be examined to find the minimum, while on the other hand $2^{m-1}-1$ partitionings ought to be evaluated. For large m the exchange method for partitions with an objective function is more economical.

It is again usually the case that various methods should be applied to the given data and distance matrices in order to select from the results obtained those most appropriate for interpretation. We ourselves and many others (Hodson 1970, Inglis and Johnson 1969, Lance and Williams 1967b, Wallace 1968) have found the KMEANS principle applied to data or distance matrices to be the most satisfactory. In contrast to the hierarchical methods where a branching or merging tree is optimally configured, in KMEANS or related methods an optimising objective function exists for any given n.

It would seem reasonable to use the results for a desired n of a hierarchical method, especially an agglomerative method, as an initial partition for an algorithm based on the KMEANS principle (see Anderberg 1973), since less computer time might be taken than is necessary with the heuristic algorithms in section 3.1 in each case. The effect seems to depend on the particular data set and on the value of m.

Chapter 5

Miscellaneous Methods

5.1 OPTIMAL PRESENTATION OF PROFILES

In this section we start with a metric data matrix

$$a_{ik} \quad (i=1,\dots,m;\ k=1,\dots,n) \tag{5.1.1}$$

which is denoted differently because it may also handle a matrix of the centroids of a partition of length m which has already been found. The number of variables is here denoted by n.

After the transformation

$$a_{ik} \longrightarrow z * \frac{a_{ik} - \alpha_i}{\beta_i - \alpha_i} \quad (k=1,\dots,n),$$
$$\alpha_i = \min_k a_{ik},\ \beta_i = \max_k a_{ik}, \quad z>0, \tag{5.1.2}$$

so that all variables vary between 0 and z, take, in a Cartesian coordinate system, the pairs of values (kh, a_{ik}) $(k = 1, \dots, n)$, where h is an arbitrary constant, e.g. $h = 1$, for each row $i = 1, \dots, m$. If the ordinates belonging to a row are then linked by straight line segments, the result is called the profile of the row. The upper half of Fig. B44 shows the profiles for six variables A, B, C, D, E and F of three centroids found by use of the KMEANS algorithm. In the lower half of Fig. B44 the order of the variables is changed to D, A, C, F, E and B, which essentially allows better interpretation of the interrelationship of the profiles of the centroids, as well as the centroids themselves and their associated clusters. Since interpretation is an essential part of the raison d'être of cluster analysis, a method for ordering the variables makes a vital contribution.

A presentation of profiles is called optimal (see Späth and Gutgesell 1972) if the order of the variables is so chosen that the number of profile intersection points is minimised. (In the diagrams, those between the profiles and the lines parallel to the vertical axis corresponding to the individual variables are not counted.) For B44 (upper) this total S is

$$S(\text{A, B, C, D, E, F}) = 10$$

and for Fig. B44 (lower) it is

$$S(\text{D, A, C, F, E, B}) = 4.$$

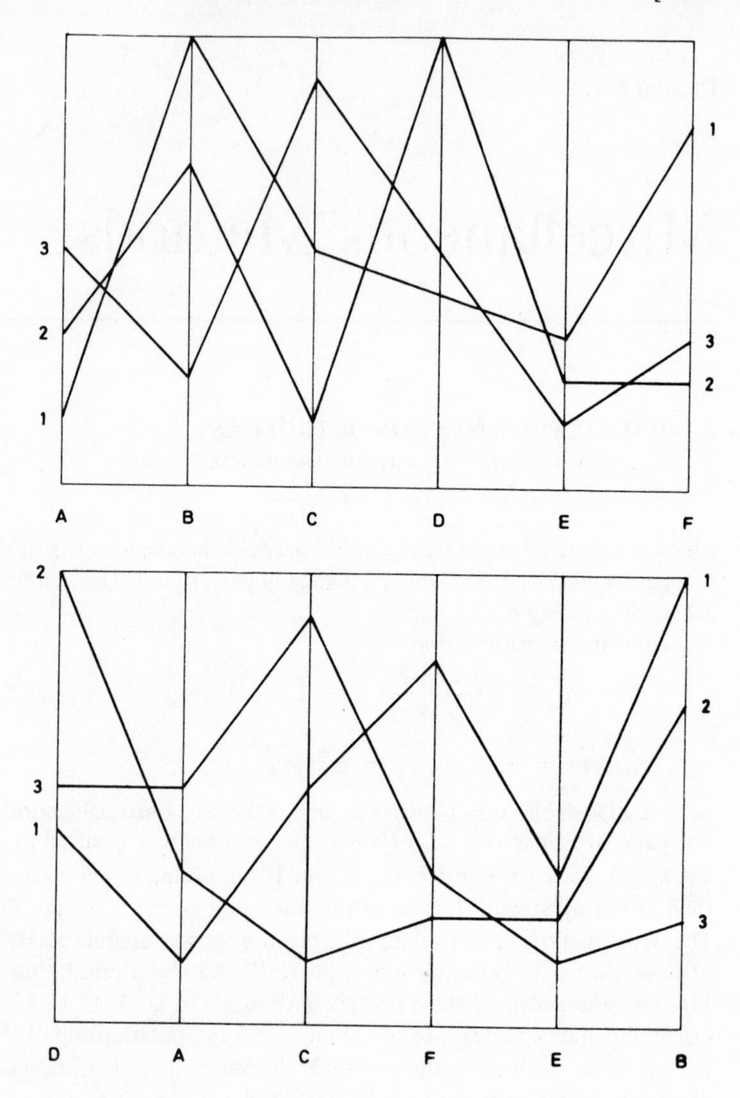

Fig. B 44

If we introduce the binary variable

$$c_{ikjl} = \begin{cases} 1 & \text{if} \quad (a_{li}-a_{ji})(a_{lk}-a_{jk})<0 \\ 0 & \text{otherwise} \end{cases} \tag{5.1.3}$$

for the condition whether or not the profiles of the jth and lth rows would intersect when the variables i and k are adjacent, then

$$d_{ik} = d_{ki} = \sum_{j=1}^{m} \sum_{l=j+1}^{m} c_{ikjl} \quad (i, k = 1, \ldots, n) \tag{5.1.4}$$

is the sum of all intersections which would be encountered between k and i or i and k.

To determine the ordering of variables which minimises the total number of intersections, a permutation $q_1, \ldots, q_n$ of the variables must be sought such that

$$S(q_1, \ldots, q_n) = d_{q_1 q_2} + d_{q_2 q_3} + \ldots + d_{q_{n-1} q_n} \tag{5.1.5}$$

is minimised. Since the matrix d_{ik} is symmetrical, we have

$$S(q_1, q_2, \ldots, q_n) = S(q_n, q_{n-1}, \ldots, q_1), \tag{5.1.6}$$

meaning that only permutations up to symmetry need to be considered.

The objective function (5.1.5) can be interpreted as a kind of travelling salesman problem. Starting from one (unknown) variable, all the others must be visited exactly once following a path which does not include the original variable and which minimises the total path length.

Fig. U 19

```
      SUBROUTINE PERML (N,X,Q,FIRST)
C
C     PERMUTATIONS OF THE ELEMENTS X(1),...,X(N) ARE GENERATED
C     IN LEXICOGRAPHIC ORDER.
C
C     Q IS AN AUXILIARY ARRAY.
C
C     THE SUBROUTINE MUST INITIALLY BE CALLED WITH
C     FIRST = .FALSE..    THE VALUE OF FIRST IS THEN SET TO .TRUE.
C     AND A PERMUTATION IS PRODUCED WITH EACH SUCCEEDING CALL.
C     ONCE ALL PERMUTATIONS ARE GENERATED, FIRST IS SET TO
C     .FALSE. AGAIN.
C
C     DIMENSION X(N), Q(N)
      DIMENSION X(20),Q(20)
      INTEGER X,Q
      LOGICAL FIRST
      N1=N-1
      IF(FIRST) GOTO 2
      FIRST=.TRUE.
      DO 1 M=1,N1
         Q(M)=N
    1 CONTINUE
    2 IF(Q(N1).NE.N) GOTO 3
      Q(N1)=N1
      L=X(N)
      X(N)=X(N1)
      X(N1)=L
      GOTO 8
```

```
    3 DO 4 J=1,N1
            K=N-J
            IF(Q(K).NE.K) GOTO 5
            Q(K)=N
    4 CONTINUE
      FIRST=.FALSE.
      K=1
      GOTO 6
    5 M=Q(K)
      L=X(M)
      X(M)=X(K)
      X(K)=L
      Q(K)=M-1
      K=K+1
    6 M=N
    7 L=X(M)
      X(M)=X(K)
      X(K)=L
      M=M-1
      K=K+1
      IF(K.LT.M) GOTO 7
    8 RETURN
      END
```

Fig. U19

```
      SUBROUTINE PERMS (N,X,FIRST,L)
C
C     WITH THE HELP OF PERML THE FIRST (N FACTORIAL)/2
C     LEXICOGRAPHIC PERMUTATIONS ARE PRODUCED.   THE
C     REMAINING (N FACTORIAL)/2 PERMUTATIONS ARE OBTAINED
C     BY INVERSION FROM THOSE PRODUCED FIRST.
C
C     THE SUBROUTINE MUST INITIALLY BE CALLED WITH
C     FIRST = .FALSE..    IT THEN SETS FIRST TO .TRUE., AND
C     REPEATED CALLS WITH X AND L UNCHANGED PRODUCE NEW
C     PERMUTATIONS.    WHEN ALL (N FACTORIAL)/2 PERMUTATIONS
C     HAVE BEEN PRODUCED, FIRST IS SET TO .FALSE. AGAIN.
C
C     DIMENSION X(N), Q(N)
      DIMENSION X(20),Q(20)
      INTEGER X,Q
      LOGICAL FIRST
      IF(FIRST) GOTO 2
      L=1
      DO 1 I=3,N
         L=L*I
    1 CONTINUE
    2 L=L-1
      IF(L.EQ.0) GOTO 3
      CALL PERML (N,X,Q,FIRST)
      GOTO 4
    3 FIRST=.FALSE.
    4 RETURN
      END
```

Fig. U 20

Since the number of variables in cluster analysis is normally not very high, or has been reduced by earlier factor analysis, full enumeration can be used without hesitation in order to minimise (5.1.5).

Subroutine PERML (Fig. U19), which is a Fortran version of the Algol procedure in (Ord-Smith), can be used to produce all permutations of n numbers in lexicographic order. The subroutine PERMS (Fig. U20) ensures that, when using PERML, only the first $n!/2$ permutations are produced, since the second $n!/2$, which are symmetrical with respect to the first, are not needed for our purpose.

In subroutine PROFIL (Fig. U21) the objective function (5.1.5) is evaluated for the $n!/2$ permutations and the permutation for which (5.1.5) is a minimum is written into the array IND. Since the binary variables (5.1.3), and hence S, are invariant with respect to transformations of the kind given in (5.1.2), the matrix (5.1.1) may be transformed beforehand or not.

Fig. U 21

```
      SUBROUTINE PROFIL (M,N,A,B,TRANS,Z,IND)
C
C     FOR A GIVEN MATRIX A, A DISTANCE MATRIX B IS COMPUTED
C     FOR ALL DIFFERENT ORDERINGS OF THE VARIABLES.
C     AN OPTIMAL ORDERING IS ASSIGNED TO THE ARRAY IND.
C     FOR TRANS = .TRUE., A IS TRANSFORMED SO THAT
C     0 <= A(I,K) <= Z FOR ALL I AND K.    IF THIS IS
C     NOT REQUIRED, TRANS SHOULD BE SET TO .FALSE.
C
C     DIMENSION A(M,N),   B(N,N),   IND(N), U(N)
      DIMENSION A(20,10),B(10,10),IND(10),U(10)
      INTEGER B,U,H,G
      LOGICAL TRANS,F
      IF(.NOT.TRANS) GOTO 4
      DO 3 I=1,N
         V=0.
         W=1.E30
         DO 1 K=1,M
            C=A(K,I)
            V=AMAX1(V,C)
            W=AMIN1(W,C)
    1    CONTINUE
         C=Z/(V-W)
         DO 2 K=1,M
            A(K,I)=C*(A(K,I)-W)
    2    CONTINUE
    3 CONTINUE
    4 N1=N-1
      M1=M-1
      B(N,N)=0
      DO 8 K=1,N1
         B(K,K)=0
         K1=K+1
         DO 7 I=K1,N
            H=0
```

```
                    DO 6 L=1,M1
                         L1=L+1
                         V=A(L,I)
                         W=A(L,K)
                         DO 5 J=L1,M
                              IF((V-A(J,I))*(W-A(J,K)).LT.0.) H=H+1
  5                      CONTINUE
  6                 CONTINUE
                    B(I,K)=H
                    B(K,I)=H
  7            CONTINUE
  8 CONTINUE
    F=.FALSE.
    DO 9 K=1,N
         U(K)=K
  9 CONTINUE
    G=99999999
 10 H=0
    J=U(1)
    DO 11 K=2,N
          I=U(K)
          H=H+B(J,I)
          J=I
 11 CONTINUE
    IF(H.GT.G) GOTO 13
    G=H
    DO 12 K=1,N
          IND(K)=U(K)
 12 CONTINUE
 13 CALL PERMS (N,U,F,L)
    IF(F) GOTO 10
    RETURN
    END
```

Fig. U 21

The main program H14 reads and prints A(M,N) in rows and, as a result of calling PROFIL, prints the matrix A(M,N), transformed if necessary with H = z as in (5.1.2), the matrix L(N,N) = (d_{ik}) and the permutation IR(N) which has been found.

Fig. H 14

```
C
C       OPTIMAL PRESENTATION OF PROFILES
C
        DIMENSION A(20,10),L(10,10),IR(10)
        LOGICAL TRANS
        KI=5
        KO=6
        TRANS=.TRUE.
      1 READ(KI,2) M,N,H
      2 FORMAT(2I5,F5.0)
        IF(M.LE.0.OR.M.GT.20.OR.N.LE.0.OR.N.GT.20) STOP
        WRITE(KO,3)
```

```
    3 FORMAT('1')
      DO 4 I=1,M
            READ(KI,5)  (A(I,K),K=1,N)
            WRITE(KO,6) (A(I,K),K=1,N)
    4 CONTINUE
    5 FORMAT(16F5.0)
    6 FORMAT(10X,10F10.3)
C
      CALL PROFIL (M,N,A,L,TRANS,H,IR)
C
      WRITE(KO,7)
    7 FORMAT('0')
      DO 8 I=1,M
            WRITE(KO,6) (A(I,K),K=1,N)
    8 CONTINUE
      WRITE(KO,7)
      DO 9 I=1,N
            WRITE(KO,10) (L(I,K),K=1,N)
    9 CONTINUE
   10 FORMAT(20X,10I5)
      WRITE(KO,7)
      WRITE(KO,10) (IR(K),K=1,N)
      GOTO 1
      END
```

Fig. H 14

In the example in Fig. E14.1 the values were taken of five socio-economic variables for twelve municipalities, from (Harman 1970) – hence there were no centroids of clusters. Fig. B45 shows the profiles for the assumed ordering sequence (A, B, C, D, E) = (1, 2, 3, 4, 5), and for the optimal presentation, with the ordering (B, E, D, A, C) = (C, A, D, E, B) = (3, 1, 4, 5, 2).

Fig. E 14.1

```
5700.000     12.800   2500.000     270.000  25000.000
1000.000     10.900    600.000      10.000  10000.000
3400.000      8.800   1000.000      10.000   9000.000
3800.000     13.600   1700.000     140.000  25000.000
4000.000     12.800   1600.000     140.000  25000.000
8200.000      8.300   2600.000      60.000  12000.000
1200.000     11.400    400.000      10.000  16000.000
9100.000     11.500   3300.000      60.000  14000.000
9900.000     12.500   3400.000     180.000  18000.000
9600.000     13.700   3600.000     390.000  25000.000
9600.000      9.600   3300.000      80.000  12000.000
9400.000     11.400   4000.000     100.000  13000.000


   4.225      6.667      4.667       5.474      8.000
   0.0        3.852      0.444       0.0        0.500
   2.157      0.741      1.333       0.0        0.0
   2.517      7.852      2.889       2.737      8.000
   2.697      6.667      2.667       2.737      8.000
   6.472      0.0        4.889       1.053      1.500
   0.180      4.593      0.0         0.0        3.500
   7.281      4.741      6.444       1.053      2.500
```

```
8.000      6.222      6.667      3.579      4.500
7.730      8.000      7.111      8.000      8.000
7.730      1.926      6.444      1.474      1.500
7.551      4.593      8.000      1.895      2.000

        0   28    6   15   25
       28    0   27   10    4
        6   27    0   15   25
       15   10   15    0    8
       25    4   25    8    0

        3    1    4    5    2
```

Fig. E 14.1

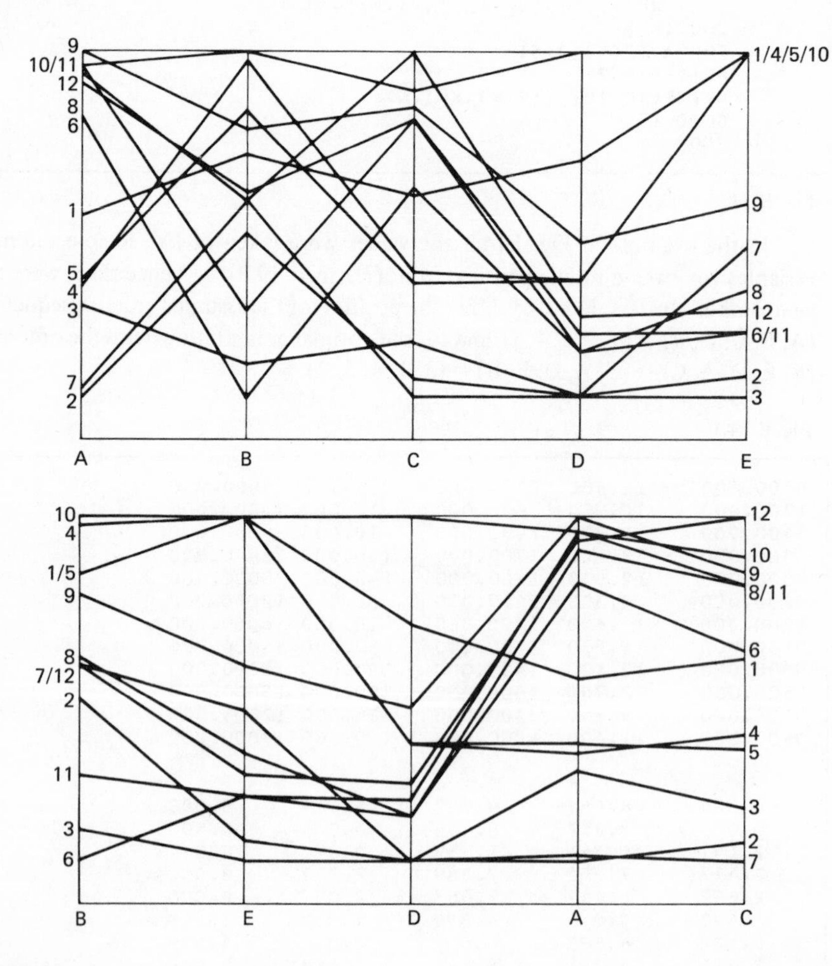

Fig. B 45

This example shows that, for small sets of m objects, optimal presentation of profiles can help not only in better interpretation of cluster centroids, but also in visual detection of clusters where there are more than two variables.

Better presentations can be achieved (Dichtl) by allowing changes of variable direction and therefore also transformations of the form

$$a_{ik} \longrightarrow z * \left(1 - \frac{a_{ik} - \alpha_i}{\beta_i - \alpha_i}\right) \tag{5.1.7}$$

in place of (5.1.2) before PROFIL is used. We have not taken this into consideration; the number of intersections would certainly be reduced, but we do not believe that interpretability is improved by changes of direction, since this may involve arbitrary changes for variables from original natural directions that are available.

5.2 THE BOND ENERGY ALGORITHM

It is assumed in this section we have either an arbitrary symmetric distance matrix, or an asymmetric matrix, corresponding to the data matrix (1.1). For simplicity let the values contained therein be already converted in such a way that, by rounding up and down, only integer, non-negative values appear, where a zero indicates the absence of a relationship for the corresponding columns and rows, and a larger integer indicates a higher degree of relationship, in proportion to the magnitude of the integer. The matrix therefore contains ordinal data.

The following algorithm (McCormick *et al.* 1969, 1972) attempts to arrange the columns and rows of the given matrix

$$a_{ik} \quad (i = 1, \ldots, m;\ k = 1, \ldots, l) \tag{5.2.1}$$

in such a way that larger elements of the matrix are nearer together and relationships between objects and objects or variables and variables, or for unsymmetrical matrices between objects and variables, are visually more obvious.

Accordingly one tries to maximise the 'bond energy'

$$M(1, \ldots, m; 1, \ldots, l) = \sum_{i=1}^{m} \sum_{k=1}^{l} a_{ik} \left[a_{i,k-1} + a_{i,k+1} + a_{i+1,k} + a_{i-1,k}\right], \tag{5.2.2}$$

in which

$$a_{i,0} = a_{i,l+1} = a_{0,k} = a_{m+1,k} = 0 \tag{5.2.3}$$

by seeking permutations $(q_1, \ldots, q_m)$ of $(1, \ldots, m)$ and $(p_1, \ldots, p_l)$ of $(1, \ldots, l)$ are such that

$$M(q_1, \ldots, q_m; p_1, \ldots, p_l) \longrightarrow \max \tag{5.2.4}$$

which is then termed the maximal bond energy.

Since (5.2.2) can be split up into

$$M = M_Z + M_S \tag{5.2.5}$$

where

$$M_S(1, \ldots, m) = \sum_{i=1}^{m} \sum_{k=1}^{l} a_{ik}(a_{i,k-1} + a_{i,k+1}) \tag{5.2.6}$$

and

$$M_Z(1, \ldots, l) = \sum_{i=1}^{m} \sum_{k=1}^{l} a_{ik}(a_{i-1,k} + a_{i+1,k}), \tag{5.2.7}$$

M can be maximised by dealing separately with M_z and M_s, which represent the sums corresponding to rows and columns. When $m = l$ and the matrix is symmetric, $M_z = M_s = M/2$.

We are able in the following to limit the discussion to the maximisation of either (5.2.6) or (5.2.7). We select M_z.

Since there are $l!$ possible permutations of the columns (in the case of M_s, $m!$ permutations of the rows), full enumeration is not usually feasible in practice. This means a return to heuristic methods for solving two travelling salesman problems (see Lenstra 1974).

For this reason (McCormick *et al.* 1972) suggest the following step-wise optimal algorithm, which possesses similarities to the KMEANS principle: for $k = 1, \ldots, l-1$, all $l - k$ columns numbered $k+1, \ldots, l$ are experimentally put in turn into each of the possible $k + 1$ positions relative to the k columns which have been dealt with. The kth column, once considered, is inserted in that position where it increases the bond energy (5.2.7) the most. The result of this algorithm depends only on the choice of the first column, so the procedure usually to be recommended is to try out each column in turn as the first, and to select the best result. Nevertheless this might not be a global optimum.

Subroutine COLPER (Fig. U22) applies this algorithm to a given permutation P(L), which is then altered accordingly. If the argument POWER is positive and not equal to 1, the elements of the matrix (5.2.1) are raised to that power before the algorithm is applied.

Fig. U22

```
      SUBROUTINE COLPER (A,M,L,P,POWER)
C
C     A PERMUTATION OF INDICES P(L) IS SOUGHT SUCH THAT,
C     FOR A GIVEN INTEGER MATRIX A(M,L) (WHICH, WITHOUT
C     LOSS OF GENERALITY, HAS NON-NEGATIVE VALUES), THE SUM
C
C     SUM(SUM(B(I,P(K))*(B(I,P(K-1))+B(I,P(K+1)))))
C      I   K
C
C     WHERE B(I,K) = A(I,K)**POWER, IS MAXIMISED.
C     B(I,0) = B(I,L+1) = 0 IN THIS FORMULA.
C
C     FOR THIS PURPOSE, FOR K = 1,...,L-1, ALL L-K COLUMNS
C     K+1,...,L ARE PLACED EXPERIMENTALLY IN EACH OF THE
C     K+1 POSSIBLE POSITIONS RELATIVE TO THE FIRST K COLUMNS
C     WHICH HAVE ALREADY BEEN DEALT WITH.   THE (K+1)-TH COLUMN
C     IS THEN INSERTED WHERE IT MOST INCREASES THE BOND ENERGY.
C
      INTEGER A,P,Q,R,S,T,U,V
C     DIMENSION A(M,L),  B(M,L),  P(L)
      DIMENSION A(60,60),B(60,60),P(60)
      DO 2 K=1,L
            DO 1 I=1,M
                  B(I,K)=(FLOAT(A(I,K)))**POWER
    1       CONTINUE
    2 CONTINUE
      L1=L-1
      DO 10 K=1,L1
            H=0.
            K1=K+1
            DO 8 U=K1,L
                  Q=P(U)
                  DO 7 V=1,K1
                        G=0.
                        IF(V.EQ.1) GOTO 4
                        N=P(V-1)
                        DO 3 I=1,M
                              G=G+B(I,Q)*B(I,N)
    3                   CONTINUE
    4                   IF(V.EQ.L) GOTO 6
                        N=P(V)
                        DO 5 I=1,M
                              G=G+B(I,Q)*B(I,N)
    5                   CONTINUE
    6                   IF(G.LE.H) GOTO 7
                        H=G
                        R=V
                        S=U
    7             CONTINUE
    8       CONTINUE
            IF(R.EQ.S) GOTO 10
            T=P(S)
            Q=R+1
```

```
          DO 9 N=Q,S
               U=S-N+Q-1
               V=U+1
               P(V)=P(U)
  9       CONTINUE
          P(R)=T
 10 CONTINUE
    RETURN
    END
```

Fig. U 22

Main program H15 solves the problem (5.2.7) for POWER = PCOL in the case of symmetry (ISYM = 0, L = M), and for ISYM≠0 solves the problem (5.2.6) for POWER = PROW, followed by the problem (5.2.7). In the second case the given matrix is transposed in advance. If ISTART = 0, the natural orderings COL(K) = K (K = 1, . . . ,L) and, if needed, ROW(I) = (I = 1, . . . ,M) are taken. If ISTART ≠ 0, initial permutations are read in; ROW(M) of course drops out if ISYM = 0.

Fig. H 15

```
C
C     SIMULTANEOUS CLASSIFICATION OF OBJECTS AND VARIABLES
C     (ROWS AND COLUMNS).
C
C     STARTING WITH A MATRIX A(M,L), WHEN M = L AND THE MATRIX IS
C     SYMMETRIC A PERMUTATION COL(L) IS SOUGHT SUCH THAT THE
C     OBJECTIVE FUNCTION IS MAXIMISED BY COLPER, WITH
C     POWER = PCOL.    OTHERWISE, THROUGH SUCCESSIVE CALLS OF
C     COLPER, FIRST WITH A(M,L) AND POWER = PCOL, AND THEN WITH
C     THE TRANSPOSED MATRIX AT(L,M) DERIVED FROM A(M,L) AND WITH
C     POWER = PROW, TWO PERMUTATIONS COL(L) AND ROW (M) ARE
C     SOUGHT SUCH THAT THE 'BOND ENERGY'
C
C     SUM(SUM(A(I,K)*(A(I,K-1)+A(I,K+1)+A(I-1,K)+A(I+1,K))))
C      I   K
C
C     IS MAXIMISED.
C
      INTEGER A(60,60),AT(60,60),COL(60),ROW(60)
      LOGICAL SYM,ASM
      KI=5
      KO=6
    1 READ(KI,2) M,L,ISYM,ISTART,PCOL,PROW
```

```
    2 FORMAT(4I5,2F5.0)
      SYM=ISYM.EQ.0
      IF(M.LE.0.OR.M.GT.60.OR.L.LE.0.OR.L.GT.60.OR.
     *    (SYM.AND.M.NE.L)) STOP
      ASM=.NOT.SYM
      IF(PCOL.EQ.0.) PCOL=1.
      IF(PROW.EQ.0.) PROW=1.
      WRITE(KO,3) M,L,ISYM,ISTART,PCOL,PROW
    3 FORMAT('1','  M=',I3,'  L=',I3,'  ISYM=',I1,
     *'  ISTART=',I1,'  PCOL=',F4.2,' PROW=',F4.2)
      WRITE(KO,4) (K,K=1,L)
    4 FORMAT('0',8X,40I3/(9X,40I3))
      IF(ISTART.NE.0) GOTO 7
      DO 5 K=1,L
           COL(K)=K
    5 CONTINUE
      IF(SYM) GOTO 8
      DO 6 I=1,M
           ROW(I)=I
    6 CONTINUE
      GOTO 8
    7 READ(KI,16) (COL(K),K=1,L)
      IF(ASM) READ(KI,16) (ROW(I),I=1,M)
    8 WRITE(KO,4) (COL(K),K=1,L)
      WRITE(KO,15)
      DO 9 I=1,M
           READ(KI,10)        (A(I,K),K=1,L)
           IF(SYM) N=COL(I)
           IF(ASM) N=ROW(I)
           WRITE(KO,11) I,N,(A(I,K),K=1,L)
    9 CONTINUE
   10 FORMAT(80I1)
   11 FORMAT(1X,I2,2X,I2,2X,40I3/(9X,40I3))
C
      CALL COLPER (A,M,L,COL,PCOL)
C
      IF(SYM) GOTO 14
      DO 13 I=1,M
            DO 12 K=1,L
                  AT(K,I)=A(I,K)
   12       CONTINUE
   13 CONTINUE
C
      CALL COLPER (AT,L,M,ROW,PROW)
C
   14 WRITE(KO,18)
      WRITE(KO,4) (K,K=1,L)
      WRITE(KO,4) (COL(K),K=1,L)
      WRITE(KO,15)
   15 FORMAT('0')
   16 FORMAT(40I2)
      DO 17 I=1,M
            IF(SYM) N=COL(I)
            IF(ASM) N=ROW(I)
            WRITE(KO,11) I,N,(A(N,COL(K)),K=1,L)
   17 CONTINUE
   18 FORMAT('1')
      GOTO 1
      END
```

Fig. H 15

The two following examples come from (McCormick *et al.* 1972) and involve symmetrical matrices. The first example (Fig. E15.1) is trivial, but the formation of blocks, which is required and obtained, can be seen from it.

```
 M=  4   L=  4   ISYM=0   ISTART=0   PCOL=1.00 PROW=1.00

             1  2  3  4

             1  2  3  4


1    1       1  0  1  0
2    2       0  1  0  1
3    3       1  0  1  0
4    4       0  1  0  1


             1  2  3  4

             3  1  4  2


1    3       1  1  0  0
2    1       1  1  0  0
3    4       0  0  1  1
4    2       0  0  1  1
```

Fig. E 15.1

The second example (Fig. E15.2) shows a matrix whose elements a_{ik} have values 0, 1, 2, and 3 for zero, weak, moderate and strong interaction between pairs i and k from 27 functional, physically separated airport buildings. Fig. E15.2 contains the rearranged matrix for ISTART = 0 and Fig. E15.3 that for ISTART$\neq$0 and COL = (18, 15, 17, 6, 23, 27, 20, 24, 26, 21, 5, 8, 10, 9, 7, 16, 15, 1, 22, 13, 2, 4, 3, 14, 19, 11, 12). The desired block formation can be seen in both cases (POWER = 1); it is somewhat better in the second case (the lower half of the table). Efficient disposition (clustering) of airport facilities can be achieved in this way.

M= 27 L= 27 ISYM=0 ISTART=0 PCOL=1.00 PROW=1.00

		1	2	3	4	5	6	7	8	9	10	11	12	13	14	15	16	17	18	19	20	21	22	23	24	25	26	27
		1	2	3	4	5	6	7	8	9	10	11	12	13	14	15	16	17	18	19	20	21	22	23	24	25	26	27
1	1	3	3	0	2	2	0	1	1	0	0	0	0	2	0	0	0	0	1	0	0	0	3	0	0	0	0	0
2	2	3	3	0	3	0	0	1	1	0	0	0	0	2	0	0	1	0	0	1	0	0	3	0	0	0	0	0
3	3	0	0	3	3	1	0	1	1	0	0	0	0	0	3	0	1	0	0	1	0	0	0	0	0	0	0	0
4	4	2	3	3	3	0	0	0	0	0	0	0	0	1	1	0	0	1	0	0	0	0	0	0	0	0	0	0
5	5	2	0	1	0	3	0	2	2	1	1	0	0	0	0	0	3	0	0	1	0	3	1	0	0	0	0	0
6	6	0	0	0	0	0	3	0	0	2	1	0	0	0	0	0	0	1	0	0	1	0	0	3	0	0	1	2
7	7	1	1	1	0	2	0	3	2	2	3	1	0	1	1	0	1	0	0	0	0	0	1	0	0	0	0	0
8	8	1	1	1	0	2	0	2	3	2	2	0	0	0	0	0	1	0	0	0	0	0	0	0	0	0	0	0
9	9	0	0	0	0	1	2	2	2	3	3	0	0	0	0	0	0	0	0	0	0	0	0	0	0	0	0	0
10	10	0	0	0	0	1	1	3	2	3	3	1	1	2	2	0	0	0	0	1	0	0	0	0	0	0	0	0
11	11	0	0	0	0	0	0	1	0	0	1	3	3	0	0	0	0	0	0	0	1	0	0	0	0	1	0	0
12	12	0	0	0	0	0	0	0	0	0	1	3	3	1	1	0	0	0	0	1	0	0	0	0	0	0	0	0
13	13	2	2	0	1	0	0	1	0	0	2	0	1	3	0	0	1	0	0	1	0	0	2	0	0	0	0	0
14	14	0	0	3	1	0	0	1	0	0	2	0	1	0	3	0	1	0	0	1	0	0	0	0	0	0	0	0
15	15	0	0	0	0	0	0	0	0	0	0	0	0	0	0	3	1	1	1	0	0	3	1	0	0	0	0	1
16	16	0	1	1	0	3	0	1	1	0	0	0	0	1	1	1	3	1	1	1	0	3	2	0	0	0	0	0
17	17	0	0	0	1	0	1	0	0	0	0	0	0	0	0	1	1	3	0	0	1	1	0	1	0	0	0	3
18	18	1	0	0	0	0	0	0	0	0	0	0	0	0	0	1	1	0	3	0	0	0	1	0	0	2	0	0
19	19	0	1	1	0	1	0	0	0	0	1	0	1	1	1	0	1	0	0	3	0	0	1	0	0	0	0	0
20	20	0	0	0	0	0	1	0	0	0	0	1	0	0	0	0	0	1	0	0	3	2	0	0	3	0	1	0
21	21	0	0	0	0	3	0	0	0	0	0	0	0	0	0	3	3	1	0	0	2	3	0	0	1	0	1	1
22	22	3	3	0	0	1	0	1	0	0	0	0	0	2	0	1	2	0	1	1	0	0	3	0	0	0	0	0
23	23	0	0	0	0	0	3	0	0	0	0	0	0	0	0	0	0	1	0	0	0	0	0	3	0	0	0	1
24	24	0	0	0	0	0	0	0	0	0	0	0	0	0	0	0	0	0	0	0	3	1	0	0	3	0	2	1
25	25	0	0	0	0	0	0	0	0	0	0	1	0	0	0	0	0	0	2	0	0	0	0	0	0	3	0	3
26	26	0	0	0	0	0	1	0	0	0	0	0	0	0	0	0	0	0	0	0	1	1	0	0	2	0	3	0
27	27	0	0	0	0	0	2	0	0	0	0	0	0	0	0	1	0	3	0	0	0	1	0	1	1	3	0	3

Fig. E 15.2

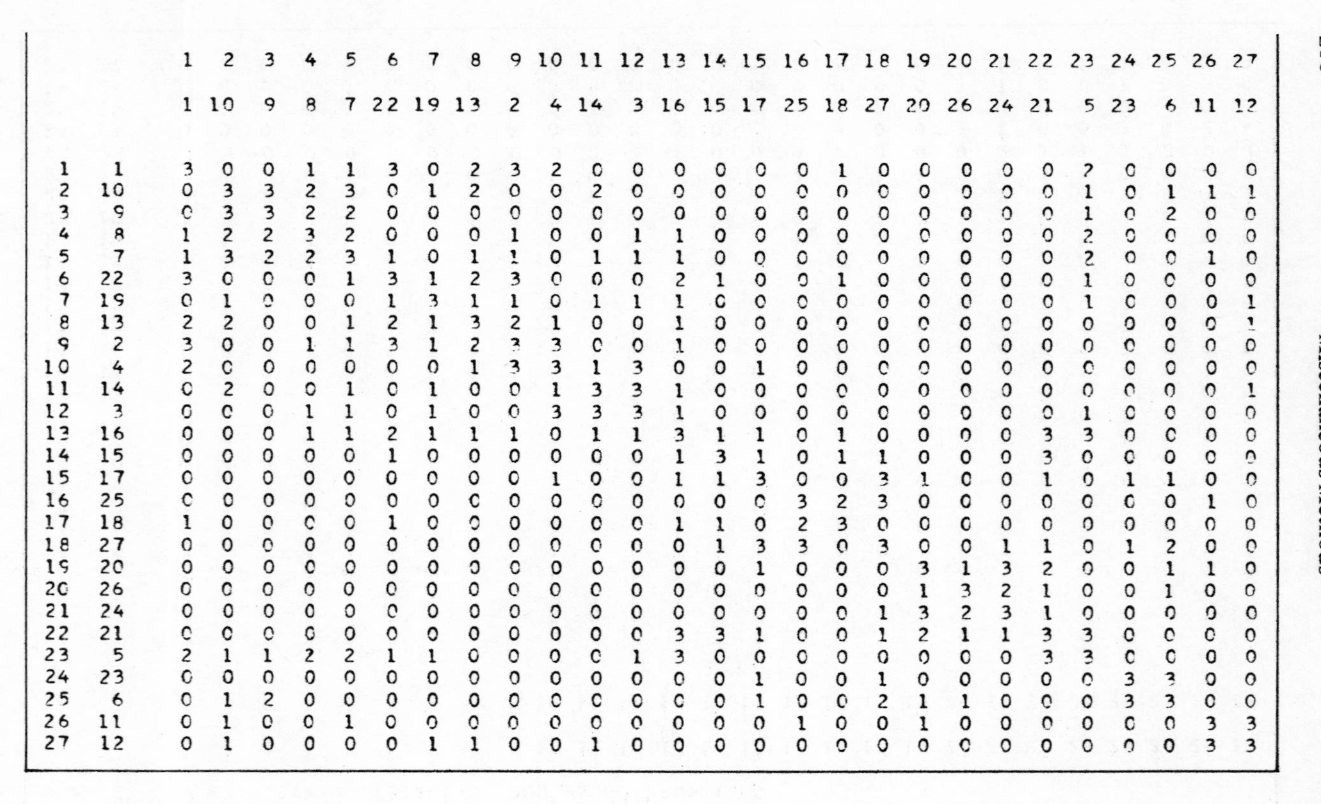

		1	2	3	4	5	6	7	8	9	10	11	12	13	14	15	16	17	18	19	20	21	22	23	24	25	26	27
		1	10	9	8	7	22	19	13	2	4	14	3	16	15	17	25	18	27	20	26	24	21	5	23	6	11	12
1	1	3	0	0	1	1	3	0	2	3	2	0	0	0	0	0	0	1	0	0	0	0	0	2	0	0	0	0
2	10	0	3	3	2	3	0	1	2	0	0	2	0	0	0	0	0	0	0	0	0	0	0	1	0	1	1	1
3	9	0	3	3	2	2	0	0	0	0	0	0	0	0	0	0	0	0	0	0	0	0	0	1	0	2	0	0
4	8	1	2	2	3	2	0	0	0	1	0	0	1	1	0	0	0	0	0	0	0	0	0	2	0	0	0	0
5	7	1	3	2	2	3	1	0	1	1	0	1	1	1	0	0	0	0	0	0	0	0	0	2	0	0	1	0
6	22	3	0	0	0	1	3	1	2	3	0	0	0	2	1	0	0	1	0	0	0	0	0	1	0	0	0	0
7	19	0	1	0	0	0	1	3	1	1	0	1	1	1	0	0	0	0	0	0	0	0	0	1	0	0	0	1
8	13	2	2	0	0	1	2	1	3	2	1	0	0	1	0	0	0	0	0	0	0	0	0	0	0	0	0	1
9	2	3	0	0	1	1	3	1	2	3	3	0	0	1	0	0	0	0	0	0	0	0	0	0	0	0	0	0
10	4	2	0	0	0	0	0	0	1	3	3	1	3	0	0	1	0	0	0	0	0	0	0	0	0	0	0	0
11	14	0	2	0	0	1	0	1	0	0	1	3	3	1	0	0	0	0	0	0	0	0	0	0	0	0	0	1
12	3	0	0	0	1	1	0	1	0	0	3	3	3	1	0	0	0	0	0	0	0	0	0	1	0	0	0	0
13	16	0	0	0	1	1	2	1	1	1	0	1	1	3	1	1	0	1	0	0	0	0	3	3	0	0	0	0
14	15	0	0	0	0	0	1	0	0	0	0	0	0	1	3	1	0	1	1	0	0	0	3	0	0	0	0	0
15	17	0	0	0	0	0	0	0	0	0	1	0	0	1	1	3	0	0	3	1	0	0	1	0	1	1	0	0
16	25	0	0	0	0	0	0	0	0	0	0	0	0	0	0	0	3	2	3	0	0	0	0	0	0	0	1	0
17	18	1	0	0	0	0	1	0	0	0	0	0	0	1	1	0	2	3	0	0	0	0	0	0	0	0	0	0
18	27	0	0	0	0	0	0	0	0	0	0	0	0	1	3	3	0	3	0	0	1	1	0	1	2	0	0	0
19	20	0	0	0	0	0	0	0	0	0	0	0	0	0	1	0	0	0	3	1	3	2	0	0	1	1	0	0
20	26	0	0	0	0	0	0	0	0	0	0	0	0	0	0	0	0	0	1	3	2	1	0	0	1	0	0	0
21	24	0	0	0	0	0	0	0	0	0	0	0	0	0	0	0	0	1	3	2	3	1	0	0	0	0	0	0
22	21	0	0	0	0	0	0	0	0	0	0	0	3	3	1	0	0	1	2	1	1	3	3	0	0	0	0	0
23	5	2	1	1	2	2	1	1	0	0	0	0	1	3	0	0	0	0	0	0	0	3	3	0	0	0	0	0
24	23	0	0	0	0	0	0	0	0	0	0	0	0	0	1	0	0	1	0	0	0	0	0	3	3	0	0	0
25	6	0	1	2	0	0	0	0	0	0	0	0	0	0	0	1	0	0	2	1	1	0	0	0	3	3	0	0
26	11	0	1	0	0	1	0	0	0	0	0	0	0	0	0	0	1	0	0	1	0	0	0	0	0	0	3	3
27	12	0	1	0	0	0	0	1	1	0	0	1	0	0	0	0	0	0	0	0	0	0	0	0	0	0	3	3

Fig. E 15.2

M= 27 L= 27 ISYM=0 ISTART=1 PCOL=1.00 PROW=1.00

		1	2	3	4	5	6	7	8	9	10	11	12	13	14	15	16	17	18	19	20	21	22	23	24	25	26	27
		18	25	27	20	26	24	21	15	5	8	10	9	7	1	4	14	3	2	13	22	19	16	17	6	23	11	12
1	18	3	3	0	2	2	0	1	1	0	0	0	0	2	0	0	0	0	1	0	0	0	3	0	0	0	0	0
2	25	3	3	0	3	0	0	1	1	0	0	0	0	2	0	0	1	0	0	1	0	0	3	0	0	0	0	0
3	27	0	0	3	3	1	0	1	1	0	0	0	0	0	3	0	1	0	0	1	0	0	0	0	0	0	0	0
4	20	2	3	3	3	0	0	0	0	0	0	0	0	1	1	0	0	1	0	0	0	0	0	0	0	0	0	0
5	26	2	0	1	0	3	0	2	2	1	1	0	0	0	0	0	3	0	0	1	0	3	1	0	0	0	0	0
6	24	0	0	0	0	0	3	0	0	2	1	0	0	0	0	0	0	1	0	0	1	0	0	3	0	0	1	2
7	21	1	1	1	0	2	0	3	2	2	3	1	0	1	1	0	1	0	0	0	0	0	1	0	0	0	0	0
8	15	1	1	1	0	2	0	2	3	2	2	0	0	0	0	0	1	0	0	0	0	0	0	0	0	0	0	0
9	5	0	0	0	0	1	2	2	2	3	3	0	0	0	0	0	0	0	0	0	0	0	0	0	0	0	0	0
10	8	0	0	0	0	1	1	3	2	3	3	1	1	2	2	0	0	0	0	1	0	0	0	0	0	0	0	0
11	10	0	0	0	0	0	0	1	0	0	1	3	3	0	0	0	0	0	0	0	1	0	0	0	0	1	0	0
12	9	0	0	0	0	0	0	0	0	0	1	3	3	1	1	0	0	0	0	1	0	0	0	0	0	0	0	0
13	7	2	2	0	1	0	0	1	0	0	2	0	1	3	0	0	1	0	0	1	0	0	2	0	0	0	0	0
14	1	0	0	3	1	0	0	1	0	0	2	0	1	0	3	0	1	0	0	1	0	0	0	0	0	0	0	0
15	4	0	0	0	0	0	0	0	0	0	0	0	0	0	0	3	1	1	1	0	0	3	1	0	0	0	0	1
16	14	0	1	1	0	3	0	1	1	0	0	0	0	1	1	1	3	1	1	1	0	3	2	0	0	0	0	0
17	3	0	0	0	1	0	1	0	0	0	0	0	0	0	0	1	1	3	0	0	1	1	0	1	0	0	0	3
18	2	1	0	0	0	0	0	0	0	0	0	0	0	0	0	1	1	0	3	0	0	0	1	0	0	2	0	0
19	13	0	1	1	0	1	0	0	0	0	1	0	1	1	1	0	1	0	0	3	0	0	1	0	0	0	0	0
20	22	0	0	0	0	0	1	0	0	0	0	1	0	0	0	0	0	1	0	0	3	2	0	0	3	0	1	0
21	19	0	0	0	0	3	0	0	0	0	0	0	0	0	0	3	3	1	0	0	2	3	0	0	1	0	1	1
22	16	3	3	0	0	1	0	1	0	0	0	0	0	2	0	1	2	0	1	1	0	0	3	0	0	0	0	0
23	17	0	0	0	0	0	3	0	0	0	0	0	0	0	0	0	0	1	0	0	0	0	0	3	0	0	0	1
24	6	0	0	0	0	0	0	0	0	0	0	0	0	0	0	0	0	0	0	0	3	1	0	0	3	0	2	1
25	23	0	0	0	0	0	0	0	0	0	0	1	0	0	0	0	0	0	2	0	0	0	0	0	0	3	0	3
26	11	0	0	0	0	0	1	0	0	0	0	0	0	0	0	0	0	0	0	0	1	1	0	0	2	0	3	0
27	12	0	0	0	0	0	2	0	0	0	0	0	0	0	0	1	0	3	0	0	0	1	0	1	1	3	0	3

Fig. E 15.3

The great advantages of this algorithm are that, in contrast to the cluster algorithms in Chapters 3 and 4, no information of any kind is lost, and that the number of clusters does not have to be presumed; it is easily and naturally visible.

There are many ways in which it can be applied. The ordinal variable $a_{ik} = a_{ki}$ can, for instance, represent the number of people sitting on both boards of directors of two firms i and k at any given time; a_{ik} can stand for the number of treaties concluded between nations i and k, or for the number of occasions on which two Members of Parliament i and k voted the same way. Non-symmetric application is indicated if a_{ik} is interpreted as a scale of values, as, perhaps, whether a type of aircraft i is suitable for a possible purpose k, or to what extent a clustering method i is applied in an application area k.

5.3 OTHER METHODS

The books available on cluster analysis (see Anderberg 1973, Bock 1974, Crouch 1971, Duda and Hart 1973, Duran and Odell 1974, Everitt 1974, Fisher 1969 and Tryon and Bailey 1970) are so numerous and so diverse, because of its many possible applications, that we cannot hope to mention every book on the subject or refer to all the cluster algorithms which are worthy of consideration. We should like to mention in this section a few more methods which seem to us to be not unimportant and which are not as widely used as those described in the previous chapters.

For example, L_p norms $(p \neq 1)$ could be applied in (3.2.5) for the objective function Z. One might perhaps have to minimise

$$Z_1^{(1)}\ (C_1, \ldots, C_n) = \sum_{j=1}^{n} \sum_{i \in C_j} \|x_i - \bar{x}_j^{(1)}\|_1 = \sum_{j=1}^{n} \sum_{i \in C_j} \sum_{k=1}^{l} |x_{ik} - \bar{x}_{jk}^{(1)}| \tag{5.3.1}$$

for $p = 1$, which has been done in section 3.5, and

$$Z_1^{(\infty)}\ (C_1, \ldots, C_n) = \sum_{j=1}^{n} \sum_{i \in C_j} \|x_i - \bar{x}_j^{(\infty)}\|_\infty = \sum_{j=1}^{n} \sum_{i \in C_j} \max_k |x_{ik} - \bar{x}_{jk}^{(\infty)}| \tag{5.3.2}$$

for $p = \infty$, in which $\bar{x}_j^{(1)}$ are the medians and $\bar{x}_j^{\infty}$ are the midpoints of the range, i.e.

$$\bar{x}_j^{(\infty)} = \tfrac{1}{2} \left(\max_{i \in C_j} x_i + \min_{i \in C_j} x_i \right)$$

The objective function $Z_1^{(1)}$ is discussed in detail in (Späth 1976) and summarised in section 3.5 ;the objective function $Z_1^{(\infty)}$ has shown itself on examination to be almost unusable, which may well be through taking outliers in the data too strongly into account (see Späth 1974). In both cases the KMEANS principle can be applied and exchange formulae can be given, like (3.2.17) and (3.2.21), which save computing time. The objective function

$$Z_1^{(\min)}(C_1, \dots, C_n) = \max_{j=1,\dots,n} \sum_{i \in C_j} \|x_i - \bar{x}_j^{(1)}\|_1 \longrightarrow \textbf{Minimum} \qquad (5.3.3)$$

might be especially meaningful in logistics.

Extending the principle of maximum centroid distance from two clusters to n clusters suggests the objective function

$$\min \max_{i \neq j} \|\bar{x}_i - \bar{x}_j\|_2 \quad (i, j = 1, \dots, n), \qquad (5.3.4)$$

(see Rao 1971), for which, however, no algorithm similar to the KMEANS principle is yet known and which, up to the present, can only be optimised through full enumeration, which is not feasible in practice. No progress has yet been made in formulating (5.3.4) in terms of linear programming (see Rao 1971).

One method which is very simple, though time-consuming computationally, is to embed the convex hull of the row vectors of a metric data matrix in an l-dimensional cube, and to examine the systematically-formed subcubes according to the number of row vectors that appear in it (see Frank and Green 1968).

The method for metric data matrices developed in (Schnell 1964) is similar; it is specified in detail in (Bock 1970) and is used extensively in (Hill *et al* 1975). The basic principle is that each vector x_i is assigned an influence on points x in its neighbourhood by means of the normal distribution, as

$$f(x; x_i; \sigma) = \frac{1}{(\sqrt{2\pi}\,\sigma)^l} \exp\left(-\frac{1}{2\sigma^2}\|x - x_i\|^2\right) \qquad (5.3.5)$$

The sum

$$F(x; x_1, \dots, x_m, \sigma) = \sum_{i=1}^{m} f(x; x_i; \sigma) \qquad (5.3.6)$$

then gives the total influence on any given vector x. The more points that lie in the neighbourhood of x, the greater F is; accordingly F is a measure of the concentrations of the x_i in the neighbourhood of x. The given x_i are now relocated by means of a very time-consuming gradient procedure in such a way that F increases from step to step. All x_i points, which are relocated in this way to the same maximum of F, form a cluster. The number of clusters n suggests itself naturally, but depends on the choice of the value of σ for the standard deviation, which can be assumed to be dependent on the points, i.e. $f(x; x_i; \sigma_i)$ instead of $f(x; x_i; \sigma)$. In our experience, considerable numerical difficulties arise during this method, which additionally requires a great deal of calculation.

A method which is essentially a generalisation of the bond energy algorithm of section 5.2 is described in (Hartigan 1968). The rows and columns of the data matrix are considered simultaneously whereby rows and columns are treated by choice using either partitioning or divisive hierarchical methods. The objective functions used are other than (5.2.2).

A whole series of methods is based on graph theoretic ideas and on the concept of maximal complete subgraphs (see Bonner 1964). By assuming a threshold value, a given distance matrix is transformed by

$$\begin{aligned} d_{ik} \geqq \tau &\longrightarrow d_{ik} = 1 \\ d_{ik} < \tau &\longrightarrow d_{ik} = 0 \end{aligned} \tag{5.3.7}$$

into a binary matrix, which is then interpreted as the incidence matrix of a graph from which the method can then be constructed (see Auguston and Minker 1970, Constantinescu 1967, Jardine and Sibson 1971). These methods are mainly used for information retrieval (see Swanson 1973, Vaswani 1969), where the binary matrix is frequently already provided and where the overlapping clusters are wanted directly.

An important so-called 'monothetic' divisive method is the AID method (see Anderberg 1973, Armstrong and Andress 1970, Morgan and Sonquist 1963, Sonquist and Morgan 1964), in which clusters are formed by considering only one variable at a time. At each step in the construction of a branching tree similar to that in DISMEA in section 4.1, ORDERD is applied for each variable and one is selected. A similar method is mentioned in (Freitag 1972). A 'monothetic' method specially suitable for mixed data is described in (Lance and Williams 1971).

We emphasise once more that we consider the most important objective functions to be (3.2.5), (3.3.16), (3.4.1), and occasionally (3.5.1) or (5.3.3) for metric data matrices, (3.5.1) being suitable for ordinal data also; and (3.2.33) for distance matrices not necessarily based on the Euclidean metric and for algorithms which partition according to the KMEANS principle. In data processing terms, apart from (3.2.33) and possibly (5.3.3), these objective functions are very suitable for sequential processing. In (Späth and Müller 1979) an exchange method for the mentioned principle of cluster-wise variance matrices is given.

Bibliography

The publications marked with an asterisk contain Algol or Fortran programs. Comprehensive reading lists are to be found in (Bock 1974) and in (Duran & Odell).

Aaker, D. A. (1971) *Multivariate Analysis in Marketing: Theory and Application.* Wadsworth Publ. Comp.

Anderberg, M. R. (1973) *Cluster Analysis for Applications.* Academic Press.

Armstrong, J. S., Andress, J. G. (1970) *Exploratory Analysis of Marketing Data: Trees vs. Regression.* J. Marketing Research **7**, 487-492.

Augustson, J. G. Minker, J. (1970) *An Analysis of Some Graph Theoretic Cluster Techniques,* J. ACM **17**, 571-588.

Ball, G. H., Hall, D. J. (1967) *A Clustering Technique for Summarizing Multivariate Data.* Behav. Science **12**, 153-155.

Ball, G. H. (1965) *Data Analysis in the Social Sciences: What about Details.* Proc. Fall Joint Comp. Conf. **27**, 533-559.

Bavarian Bureau of Statistics: Data from the 1970 population Census (in German).

Belschner, W., Späth, H. (1977) *Versuch einer Kategorisierung von erzieherischen Situationsdefinitionen mittels Cluster-Analyse.* Psychol. in Erz. u. Uhterricht **24**, 49-53.

Bijnen, E. J. (1973) *Cluster Analysis – Survey and Evaluation of Techniques.* Tilburg University Press.

Bloech, J. (1970) *Optimale Industriestandorte.* Physica-Verlag, Würzburg.

Bock, H. H. (1970) *Automatische Klassifikation.* In *Statistische Methoden II, Mehrvariable Methoden und Datenverarbeitung, Lecture Notes in Operations Research and Mathematical Systems* Nr. **39**, Springer-Verlag.

Bock, H. H. (1974) *Automatische Klassifikation.* Vandenhoeck & Rupprecht, Göttingen.

Bonner, R. E. (1964) *On Some Clustering Techniques.* IBM J. **8**, 22-32.

Braun, H. (*) (1978) *Strukturanalyse eines Tankstellennetzes.* In Späth, H. (Ed.): Fallstudien Operations Research, Band I, R. Oldenbourg, Munich.

Cheetham, A. H., Hazel, J. E. (1969) *Binary (Presence-Absence) Similarity Coefficients.* J. Paleontology **43,** 1130-1136.

Cole, A. J., Wishart, D. (1970) *An Improved Algorithm for the Jardine-Sibson Method of Generating Overlapping Clusters.* Comp. J. **13,** 156-163.

Constantinescu, P. (1967) *A Method of Cluster Analysis.* Brit. J. Math. Stat. Psych. **20,** 93-106.

Cooper, L. (1963) *Location-Allocation Problems.* Op. Res. **11,** 331-343.

Cooper, L. (1964) *Heuristic Methods for Location-Allocation Problems.* SIAM Rev. **6,** 37-53.

Cooper, L. (1967) *Solutions of Generalized Locational Equilibrium Models.* J. Regional Science **7,** 1-18.

Cooper, L. (1972) *The Transportation-Location Problem.* Op. Res. **20,** 94-108.

Crouch, D. B. (1971) *Cluster Analysis: Bibliography.* ACM SIGIR Forum 6 **(No. 3),** 11-14.

Cunningham, K. M., Ogilvie, J. C. (1972) *Evaluation of Hierarchical Grouping Techniques: A Preliminary Study.* Comp. J. **15,** 209-213.

Dattola, R. F., Murray, D. M. (1967) *An Experiment in Automatic Thesaurus Construction.* In Salton, G. (Ed.), *Information Storage and Retrieval.* Scientific Report **ISR-13,** Cornell Univ.

Deichsel, G. (1972) *Verfahren der automatischen Klassifikation durch Clusteranalyse und ihre Anwendung bie morphologischen Untersuchung an Amoeben.* Staatsexamensarbeit, Institut für Informatik, Universität Stuttgart.

Dichtl, E. Private Communication.

Diday, E., Govaert, G. (1974) *Classification avec distance adaptive.* C. R. Acad. Sc. Paris A, 993-995.

Duda, O., Hart, P. E. (1973) *Pattern Classification and Scene Analysis.* Wiley.

Duran, B. S., Odell, P. L. (1974) *Cluster Analysis: A Survey.* Springer-Verlag.

Edwards, A. W. F., Cavalli-Sforza, L. L. (1965) *A Method for Cluster Analysis.* Biometrics **21,** 362-275.

Elton, E. J. Gruber, M. J. (1970) *Homogeneous Groups and the Testing of Economic Hypotheses.* J. Fin. Quant. Analysis **4,** 581-602.

Elton, E. J., Gruber, M. J. (1971) *Improved Forecasting through the Design of Homogeneous Groups.* J. Bus. **44,** 432-450.

Engelman, L., Hartigan, J. A. (1969) *Percentage Points of a Test for Clusters.* J. Am. Stat. Ass. **64,** 1647-1648.

Everitt, B. (1974) *Cluster Analysis.* Heinemann Educational Books, London.

Fisher, L., Van Ness, J. W. (1971) *Admissible Clustering Procedures.* Biometrika **58,** 91-104.

Fisher, W. D. (1958) *On Grouping of Maximum Homogeneity.* J. Am. Stat. Ass. **53,** 789-798.

Fisher, W. D. (1969) *Clustering and Aggregation in Economics.* J. Hopkins Press, Baltimore.

Fleiss, J. L., Zubin, J. (1969) *On the Methods and Theory of Clustering.* Multivariate Behavioral Research **4**, 235-250.

Forsythe, G. E. Moler, C. B. (*) (1971) *Computer-Verfahren für lineare algebraische Systeme.* Oldenbourg Verlag.

Fortier, J. J. Solomon, H. (1966) *Clustering Procedures.* In Krishnahiah (Ed.), *Multivariate Analysis.* Academic Press.

Frank, R. E., Green, P. E. (1968) *Numerical Taxonomy in Marketing Analysis: A review Article.* J. Marketing Res. **5**, 83-98.

Freitag, D. (1972) *Statistische Kriterien für die Marktsegmentierung.* Jahrbuch D. Absatz-u. Verbrauchsforschung **18**, 121-128.

Friedmann, H. P., Rubin, J. (1967) *On Some Invariant Criteria for Grouping Data.* J. Am. Stat. Ass. **62**, 1159-1178.

Fritsche, M. (1973) *Automatic Clustering Techniques in Information Retrieval.* Diplomarbeit, Univ. Stuttgart.

Fukunaga, K., Koontz, W. L. G. (1970) *A Criterion and an Algorithm for Grouping Data.* IEEE Transactions on Computers **C19**, 917-923.

Goronzy, F. (1969) *A Numerical Taxonomy on Business Enterprises.* In Cole, A. J. (Ed.), *Numerical Taxonomy.* Academic Press.

Gower, J. C. (1967) *A Comparison of Some Methods of Cluster Analysis.* Biometrics **23**, 623-637.

Gower, J. C., Ross, G. J. S. (1969) *Minimum Spanning Trees and Single Linkage Cluster Analysis.* Appl. Stat. **18**, 54-64.

Green, P. E., Frank, R. E., Robinson, P. J. (1967) *Cluster Analysis in Test Market Selection.* Management Science **13**, B387-400.

Green, P. E., Rao, V. R. (1969) *A Note on Proximity Measures and Cluster Analysis.* J. Marketing Res. **6**, 359-364.

Green, P. E., Carmone, F. J. (1970) *Multidimensional Scaling and Related Techniques in Marketing Analysis.* Allyn & Bacon, Boston.

Green, P. E., Tull, D. S. (1970) *Research for Marketing Decision.* Prentice-Hall.

Green, P. E., Rao, V. R. (1972) *Applied Multidimensional Scaling.* Holt, Rinehart and Winston.

Harman, H. H. (1970) *Modern Factor Analysis.* The University of Chicago Press.

Harrison, I. (1968) *Cluster Analysis.* Metra **7**, 513-528.

Hartigan, J. A. (1968) *Direct Clustering of a Data Matrix.* J. Am. Stat. Ass. **67**, 123-129.

Hartigan, J. (*) (1975) *Clustering Algorithms.* Wiley-Interscience.

Hill, L. R. et al. (1965) *Automatic Classification of Staphylococci by Principal Component Analysis and a Gradient Method.* J. Bacteriology **89**, 1393-1401.

Hodson, F. R. (1970) *Cluster Analysis and Archaeology: Some New Developments and Applications.* World Archaeology **1**, 299-320.

Howard, R. N. (1966) *Classifying a Population into Homogeneous Groups.* In Lawrence, J. R. (Ed.), *Operational Research and Social Sciences.* Tavistock Publ., London.

Inglis, J., Johnson, D. (1969) *Some Observations on, and Developments in the Analysis of Multivariate Survey Data.* Papers of the Esomar/Apor Congress, Book I, 125-157, Amsterdam.

Jardine, N., Sibson, R. (1971) *Mathematical Taxonomy.* Wiley.

Jensen, R. E. (1969) *A Dynamic Programming Algorithm for Cluster Analysis.* J. Op. Res. Soc. Am. 7, 1034-1057.

Johnson, S. C. (1967) *Hierarchical Clustering Schemes.* Psychometrika **32**, 241-254.

Joyce, T., Channon, C. (1966) *Classifying Market Survey Respondents.* Appl. Stat. **15**, 191-215.

Kendall, M. G. (1966) *Discrimination and Classification.* In Krishnahiah (Ed.), *Multivariate Analysis.* Academic Press.

Kernan, J. B., Bruce, G. D. (1972) *The Socioeconomic Structure of an Urban Area.* J. Marketing Res. **9**, 15-18.

King, B. (1967) *Step-wise Clustering Procedures.* J. Am. Stat. Ass. **62**, 86-101.

Kruskal, J. B. (1964) *Multidimensional Scaling by Optimizing Goodness of Fit to a Nonmetric Hypothesis.* Psychometrika **29**, 1-27.

Kruskal, J. B. (1964) *Nonmetric Multidimensional Scaling: A Numerical Method.* Psychometrika **29**, 28-42.

Kuhn, H. W. (1973) *A Note on Fermat's Problem.* Math. Progr. **4**, 98-107.

Lades, R. (*) (1974) *Probleme der Cluster Analysis bei der Klassfikation bayrischer Postleitzahlgebiete nach Bevölkerungsmerkmalen.* Diplomarbeit, Universität Erlangen-Nürnberg.

Lades, R., Schläger, W. (*) (1975) Clan – *Ein EDV-Programm zur Cluster Analyse für die Rechenanlage CDC 3300.* Arbeitspapier Nr. **26**, Betriebswirtschaftliches Institut der Universität Erlangen Nürnberg.

Lance, G. N., Williams, W. T. (1967) *A General Theory of Classificatory Sorting Strategies: 1, Hierarchical Systems.* Comp. J. **9**, 373-380.

Lance, G. N., Williams, W. T, (1967) *A General Theory of Classificatory Sorting Strategies: 2. Clustering Systems.* Comp. J. **10**, 271-277.

Lance, G. N., Williams, W. T. (1967) *Mixed-Data Classificatory Programs: I. Agglomerative Systems.* Austr. Comp. J. **1**, 15-20.

Lance, G. N., Williams, W. T. (1968) *Mixed-Data Classificatory Programs: II. Divisive Systems.* Austr. Comp. J. **1**, 82-85.

Lance G. N., Milne, P. W., Williams, W. T. (1968) *Mixed-Data Classificatory Programs: III. Diagnostic Systems.* Austr. Comp. J. **1**, 178-181.

Lance, G. N., Williams, W. T. (1971) *A Note on a New Divisive Classificatory Program for Mixed Data.* Comp. J. **14**, 154-155.

Langenmayr, A., Späth, H. (1977) *Cluster-Analyse neurotischer Symptome bei Kindern.* Zeitschrift f. Klin. Psych. **6**, 83-99.

Lenstra, J. K. (1974) *Clustering a Data Array and the Travelling-Salesman Problem.* Op. Res. **22**, 413-414.

Lessig, V. P., Tollefson, J. O. (1971) *Market Segmentation Through Numerical Taxonomy.* J. Marketing Res. **8**, 480-487.

Lessig, V. P. (1972) *Comparing Cluster Analysis with Cophenetic Correlation.* J. Marketing Res. **9**, 82-84.

Ling, R. F. (1973) *A Computer Generated Aid for Cluster Analysis.* Comm. ACM **16**, 355-361.

Lüneburg, H. (1971) *Kombinatorik.* Birkhäuser – Verlag.

MacQueen, J. (1967) *Some Methods for Classification and Analysis of Multivariate Observations.* 5th Berkeley Symp. Math., Stat. Prob. **1**, 281-297.

Majone, G., Sanday, P. R. (1968) *On the Numerical Classification of Nominal Data.* Management Sciences Research Report No. **118**, Carnegie-Mellon University.

Marriot, F. H. C. (1971) *Practical Problems in a Method of Cluster Analysis.* Biometrics **27**, 501-514.

McCormick, W. T., Deutsch, S. B., Martin, J. J., Schweitzer, P. J. (1969) *Identification of Data Structures and Relationships by Matrix Reordering Techniques.* Research Paper P-**512**, Institute for Defense Analysis, Arlington, U.S.A.

McCormick, W. T., Schweitzer, P. J., White, T. W. (1972) *Problem Decomposition and Data Reorganization by a Clustering Technique.* Op Res. **20**, 993-1009.

McRae, D. J. Mikca, (1971) *A Fortran IV Iterative K-Means Cluster Analysis Program.* Behav. Sci. **16**, 423-424.

Morgan, J. N., Sonquist, J. A. (1963) *Problems in the Analysis of Survey Data, and a Proposal.* J. Amer. Stat. Ass. **58**, 415-434.

Morrison, D. G. (1967) *Measurement Problems in Cluster Analysis.* Management Science **13**, B755-780.

Ord-Smith, R. J. (*) *Generation of Permutations in Lexicographic Order.* Collected Algorithms Comm. ACM **323**.

Ortega, J. M., Rheinboldt, W. C. (1970) *Iterative Solution of Nonlinear Equations in Several Variables.* Academic Press.

Overall, J. E. (1964) *Note on Multivariate Methods for Profile Analysis.* Psych. Bull. **61**, 195-198.

Parks, J. M. (1969) *Classification of Mixed Mode Data by R-Mode Factor Analysis and Q-Mode Cluster Analysis on Distance Function.* In Cole, A. J., *Numerical Taxonomy,* Academic Press.

Rand, W. M. (1971) *Objective Criteria for the Evaluation of Clustering Methods.* J. Am. Stat. Ass. **66**, 856-850.

Rao, M. R. (1971) *Cluster Analysis and Mathematical Programming.* J. Am. Stat. Ass. **66**, 622-626.

Reitsma, K., Sagalyn, J. (1967) *Correlation Measures.* In Salton, G. (Ed.), *Information Storage and Retrieval.* Scientific Report ISR-**13**, Cornell Univ.

Restle, F. (1961) *Psychology of Judgment and Choice.* Wiley.

Rijsbergen, Van, C. J. (*) (1970) *A Fast Hierarchic Clustering Algorithm.* Comp. J. **13**, 324-326.

Rogers, D. J., Tanimoto, T. T. (1960) *A Computer Program for Classifying Plants.* Science **132,** 1115-1118.

Rohlf, F. J. (1970) *Adaptive Hierarchical Clustering Schemes.* Syst. Zoology **19,** 58-82.

Rohlf, F. J. (*) (1974) *Dendrogramm Plot.* Computer J, **17,** 89-91.

Ross, G. J. S. (*) (1969) *Algorithm AS 13: Minimum Spanning Tree.* Appl. Stat. **18,** 103-104.

Ross, G. J. S. (*) (1969) *Algorithm AS 14: Printing the Minimum Spanning Tree.* Appl. Stat. **18,** 105-106.

Ross, G. J. S. (*) (1969) *Algorithm AS 15: Single Linkage Cluster Analysis.* Appl. Stat. **18,** 106-108.

Ross, G. J. S. (1969) *Classification Techniques for Large Sets of Data.* In Cole, A. J. *Numerical Taxonomy.* Academic Press.

Rubin, J. (1967) *Optimal Classification Into Groups: An Approach for Solving the Taxonomy Problem.* J. Theor. Biol. **15,** 103-144.

Sale, A. H. S. (*) (1971) *An Improved Clustering Algorithm.* Computer J. **14,** 104-106.

Schnell, P. (1964) *Eine Methode zum Auffinden von Gruppen.* Biom. Zeitschr. **6,** 47-48.

Scott, A. J., Symons, M. J. (1971) *Clustering Methods Based on Likelihood Ratio Criteria.* Biometrics **17,** 387-397.

Seppänen, J. J. (*) *Spanning Tree.* Collected Algorithms CACM **399.**

Shepard, R. N., Romney, A. K., Nerlove, S. B. (1972) *Multidimensional Scaling – Theory and Applications in the Behavioral Sciences, Vol. I and II.* Seminar Press, New York.

Sibson, R. (*) (1973) *An Optimally Efficient Algorithm for the Single-Linkage Cluster Method.* Comp. J. **16,** 30-34.

Sneath, P. H. A. (1966) *A Method for Curve Seeking from Scattered Points.* Computer J. **8,** 383-391.

Sneath, P. H. A., Sokal, R. R. (1973) *Numerical Taxonomy.* San Francisco.

Soergel, D. (1967) *Mathematical Analysis of Documentation Systems.* Inform. Stor. Retr. **3,** 129-173.

Sokal, R. R., Sneath, P. H. A. (1963) *Principles of Numerical Taxonomy.* Freeman and Co.

Sonquist, J. A., Morgan, J. N. (1964) *The Detection of Interaction Effects.* Monograph No. **35,** Institute for Social Research, University of Michigan.

Späth, H., Gutgesell, W. (*) (1972) *Zur optimalen Darstellung von Profilen.* Angew. Informatik 575-577.

Späth, H. (*) (1973) *Clustering of One-Dimensional Ordered Data.* Computing **11,** 175-177.

Späth, H. (1973) *Bedarfsvorhersage via Cluster Analysis.* DGOR-Proceedings in Operations Research **3,** 414. Physica-Verlag, Würzburg.

Späth H. (*) (1974) *Algorithmen für multivariable Ausgleichsmodelle.* R. Oldenbourg Verlag.

Späth, H. (*) (1976) *L_1 Cluster Analysis.* Computing **16**, 379-387.

Späth, H. (1977) *Numerische Erfahrungen zu heuristischen Lösungsverfahren beim Varianzkriterium in der Cluster-Analyse.* Angew. Informatik 2/77, 67-72.

Späth, H. (1977) *Computational Experiences with the Exchange Method applied to Four Commonly Used Partitioning Cluster Analysis Criteria.* European J. of Operational Research, **1**, 23-31.

Späth, H. (1977) *Partitionierende Cluster-Analyse für große Objektemengen mit binären Merkmalen am Beispiel von Firmen und deren Berufsgruppenbedarf.* In Späth, H. (Hrsg.), *Fallstudien Cluster-Analyse.* R. Oldenbourg Verlag.

Späth, H. (1977) *Partitionierende Cluster-Analyse bei Binärdaten am Beispiel von bundesdeutschen Hochschulen und Diplomstudiengängen.* Zeitschrift für Operations Research **21**, B85-96.

Späth, H. (1977) *Numerischer Vergleich von zwei kanonischen Varianten des Austauschverfahrens beim Varianzkriterium in der Cluster-Analyse.* Angew Informatik 9/77, 295-397.

Späth, H. (1978) *Bedarfsvorhersage für saisonale Grossortimente in Handelsunternehmen.* In: Späth, H. (Ed.): Fallstudien Operations Research, Band I, R. Oldenbourg, Munich.

Späth, H., Müller, R. (1979) *Das Austausch verfahren für die skalienngsinvariante Methode der adaptiven Distanzen in der Clusteranalyse.* In Späth, H. (Ed.): *Ausgewählte Operations Research Software in Fortran,* Munich 1979.

Sparks, D. N. (*) (1973) *Euclidean Cluster Analysis.* Appl. Stat. **22**, 126-130.

Statistisches Bundesamt. (1973) *Wohnbevölkerung in den Postleiteinheiten und in ausgewählten administrativen Gebietseinheiten am 27.5.1970.* Kohlhammer-Verlag.

Swanson, R. W. (1973) *On Clustering Techniques in Information Science.* J. Am. Soc. Inf. Sci. **24**, 72-73.

Talkington, L. (1967) *A Method of Scaling for a Mixed Set of Discrete and Continuous Variables.* Syst. Zoology **16**, 149-152.

Tryon, R. C., Bailey, D. E. (1970) *Cluster Analysis.* McGraw-Hill.

Uberla, K (*) (1971) *Faktorenanalyse.* Springer-Verlag.

Vaswani, P. K. T. (1969) *A Technique for Cluster Emphasis and its Application to Automatic Indexing.* Inform. Retrieval 1300-1303.

Veldman, D. J. (*) (1967) *FORTRAN Programming for the Behavioral Sciences.* Holt, Rinehart and Winston.

Vinod, H. D. (1969) *Integer Programming and the Theory of Grouping.* J. Am. Stat. Ass. **64**, 506-519.

Wallace, D. L. (1968) *Clustering.* Int. Encycl. Soc. Sci., Crowell Collier.

Ward, J. H. (1963) *Hierarchical Grouping to Optimize an Objective Function.* J. Am. Stat. Ass. **58**, 236-244.

Index